SHELLS

FABIO MORETZSOHN

S H
E L
L S

A NATURAL AND CULTURAL HISTORY

REAKTION BOOKS

PUBLISHED BY
REAKTION BOOKS LTD
UNIT 32, WATERSIDE
44-48 WHARF ROAD
LONDON N1 7UX, UK
WWW.REAKTIONBOOKS.CO.UK

FIRST PUBLISHED 2023

PRINTED AND BOUND IN INDIA BY REPLIKA PRESS PVT. LTD

A CATALOGUE RECORD FOR THIS BOOK IS AVAILABLE
FROM THE BRITISH LIBRARY

ISBN 978 1 78914 713 1

CONTENTS

CHAPTER ONE

THE SHELL
MAKERS

Shells are the hard, protective outer cases of molluscs or crustaceans. While many unrelated organisms also secrete shells to protect their bodies, such as foraminifera, sea urchins and turtles, this book focuses on the calcareous shells secreted by snails, clams and other molluscs.

Shells have captivated humans – and their ancestors – since the dawn of time. We have recently learned that the earliest known artwork was made on a shell. Molluscs have long been collected for food, and their shells used as tools, as jewellery, to decorate their dwellings, to bring good luck or to ward off spirits. Shells were also used as symbols of fertility and in religious rituals. Several Indigenous peoples, independently, have used different species of shells as currency, and in a few places they continue to do so. We'll look at some of the most popular groups of shells, cowries and their kin, and how their beautiful and colourful shells are formed. Pearls are the only gems of animal origin; they are made by the same molluscs that produce shells. Their iridescent beauty has fascinated humans for thousands of years. Shells have inspired many artists and have been featured in all forms of art, including architecture. Finally, shells are used by scientists to provide clues about changing environmental conditions, in order to prepare us for the future.

Molluscs have a soft body (*mollis* meaning 'soft' in Latin), and generally have an external calcareous shell that protects their bodies. Although most molluscs produce a shell, some have a reduced one or have completely lost their shell (for example, nudibranch gastropods). When a shell is present, the animal is intimately connected to it and cannot leave it to look for a larger home. Old observations of gastropods leaving the shell and finding a bigger one are not true. Instead, the mollusc keeps adding more shell material to enlarge it as the animal grows. However, hermit crabs, which are crustaceans, use gastropods shells temporarily, until the crab grows too large and needs to find a bigger home.

Molluscs are invertebrates and the second most diverse phylum, with about 93,000 living species and 70,000 fossil species estimated in 2003,[1] exceeded only by arthropods in diversity. Research suggests that less than half of the living molluscs have been discovered, and in recent years about six hundred new species have been described annually. Molluscs are among the most successful organisms and have colonized nearly all habitats. On land, they occur in forests, deserts and mountain tops, as well as in rivers, lakes and brackish water. They even make brief forays into the air: some squids have been documented swimming rapidly near the water's surface, jumping out of the water and sustaining powered flight for up to 30 metres (100 ft), supposedly to escape predators.[2] Molluscs are distributed worldwide, from the Arctic to the Antarctic, from above the tide line to the bottom of the ocean, at depths deeper than the highest peak of Mount Everest.[3]

Meet the Family

Molluscs are classified in the phylum Mollusca and divided into eight classes: Bivalvia (which includes clams, scallops and oysters); Gastropoda (snails, whelks, conchs);

Hysteroconcha lupanaria, a spiny clam from western Mexico, lives buried in sand. Its long spines pointed towards the surface of the sediment protect its fleshy siphons.

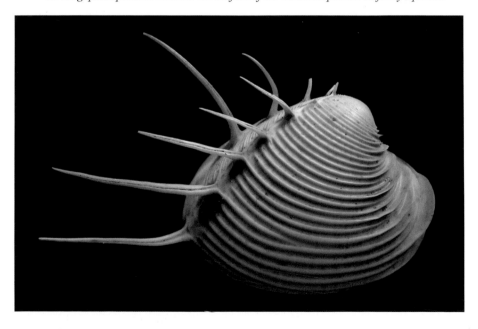

Cephalopoda (squids, octopuses and Nautilus); Caudofoveata; Solenogastres; Monoplacophora; Polyplacophora; and Scaphopoda. The most familiar molluscs are members of the first three classes.

Bivalves have a shell divided into two parts, or 'valves', which are usually similar in size and shape, but in some clams the two valves can be quite different, for example, in some scallops. The valves are connected by an elastic material, the ligament, in the hinge area. The shell has interlocking teeth near the ligament that keep the two valves together and properly aligned. The soft body of the clam is protected within its shell, and a pair of adductor muscles closes the valves when contracted. When the adductor muscles relax, the flexible ligament passively opens the shell, meaning that a clam's siphons and foot can expand and protrude. Clams are typically free-living and move around with the aid of a muscular foot that they use as an anchor or to burrow into the sediment. Species that live deeply buried within the sediment have long siphons that can extend above the surface of the sediment: for example, the geoduck, *Panopea generosa*, has a pair of siphons that can extend to a length of about 1 metre (3 ft). Some bivalves, such as oysters, do not live buried at all but rather spend their adult lives attached permanently to hard substrates; those may lack siphons. Clams, oysters and mussels are an important part of the human diet, and there are large fisheries for many species. There are about 20,000 species of living bivalves: the majority live in the oceans, and 1,200 species live primarily in fresh water, fewer in brackish water.[4]

Bivalves have a simple nervous system consisting of a network of nerves and ganglia that connect to sensory organs and control muscles. Although most bivalves lack eyes, instead utilizing simple photosensitive cells that can distinguish light and darkness to detect the presence of predators, scallops may have up to a hundred complex eyes along their mantle edge. Scallops are among the most active bivalves and can swim through the water by opening and closing their valves rapidly. This is an escape response to predatory starfish; it is not pretty, but it is effective enough to escape the immediate threat the starfish pose.

The majority of bivalves are filter feeders: water flows in through their inhalant siphon, and they use their gills to filter small food particles from it. Several bivalves,

Opposite page: Bivalve diversity in shell shape and sculpture, illustration from Ernst Haeckel, Kunstformen der Natur *(1904). Overleaf: The eight classes of living molluscs: 1) Bivalvia, 2) Gastropoda, 3) Polyplacophora, 4) Monoplacophora, 5) Scaphopoda, 6) Cephalopoda, 7) Caudofoveata and 8) Solenogastres.*

5

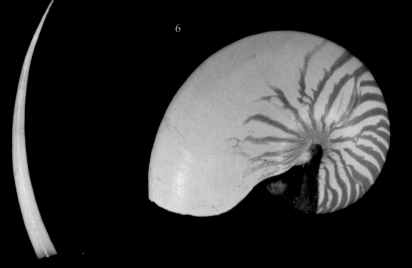

6

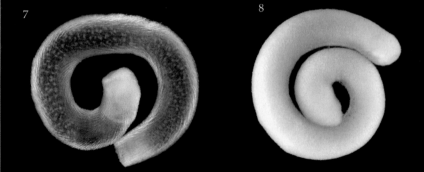

7

8

Above: Tridacna maxima *is a smaller cousin of the giant clam,* T. gigas, *but has a more brightly coloured mantle. All species of giant clams, ten in the genus* Tridacna *and two in the genus* Hippopus, *are considered at risk because of overfishing for food and the aquarium industry. Opposite: Mitchell's wentletrap (*Amaea mitchelli*) is a collector's favourite, and the symbol of the Coastal Bend Shell Club, Corpus Christi, Texas.*

such as deep-sea clams, and some shallow-water species, such as lucinids, obtain their food from symbiotic bacteria that use the energy released by chemical reactions to make food; others are deposit feeders, or even carnivores. Less common methods of feeding include bloodsucking or the sucking of body fluids in parasitic coral-boring bivalves, and the ingesting of cellulose by shipworms, which are not worms but small bivalves that bore into waterlogged wood. Shipworms have cellulose-digesting bacteria in their guts that allow them to feed on this low-energy food source. Giant clams (formerly in the family Tridacnidae, now considered a subfamily of Cardiidae[5]) are solar powered: they harbour photosynthetic zooxanthellae (algae) that confer bright colours on their mantles and convert sunlight into food for their host.

Gastropods are univalves, with a single, usually coiled shell. The more familiar species include conchs, whelks, murexes, cowries, cone shells and land snails, and others without a shell, such as nudibranchs and slugs. The typical gastropod shell is a coiled hollow tube that the animal carries on its back while it crawls using its large

muscular foot. Snail shells come in many different shapes, ranging from broadly conical to turreted or globose. Most gastropods have an operculum, which protects the body when the animal is fully withdrawn into its shell. Gastropods are the most diverse of the molluscs, with an estimated 70,000 species; most gastropods live in the ocean, but about 31,000 species live on land, and some 3,000 in fresh water.[6]

The body of gastropods consists of four main regions: a large muscular foot, which it uses to crawl; the head, with tentacles, eyes and a proboscis; the visceral hump, tucked deepest into the shell, where the digestive gland, heart, gills and other organs are located; and the mantle, which is a thin flap of tissue that secretes the shell. The pair of eyes can be located either at the base of the tentacles, as in some marine gastropods, or at the tips of the tentacles, as in land snails. Gastropod eyes range

*Below: Queen conch (*Aliger gigas*) and the large claw-like operculum that the animal uses to anchor in the sediment and jump forward like a pole-vaulter, illustration from L.-C. Kiener,* Spécies général et iconographie des coquilles vivantes, *vol. IV (1843). Opposite: Queen conch feeding on the seabed.*

from a simple eye pit, as in limpets, to complex, well-developed eyes with lenses, as in conchs. However, even in snails with the most complex eyes, vision is not the primary sense. Gastropods rely predominantly on chemosensory organs to detect chemical cues in the water or sediment not only to find mates and food, but to avoid predators.

Feeding strategies are diverse among gastropods. Many groups, such as limpets, conchs and cowries, are herbivores. These animals use their radula, a rasping organ, to scrape small pieces of tissue from algae, sea-grass blades and other plants. Some gastropods feed on sponges, bryozoans, corals and other sessile prey, while others feed on mobile prey and have to track them down by their scent; nutmegs (Cancellariidae) are highly specialized predators that suck the blood of their prey – in some cases even sharks. Mudsnails and other gastropods that feed on carrion may also use scent trails to locate decomposing animals. Parasitism has evolved in many gastropod groups, such as eulimids and pyramidellids; some simply attach to their host and suck body fluids, while others have a reduced shell and live within the tissues of their host. The slime trail left by land snails and marine gastropods can also be used by carnivorous snails and other predators to locate their snail prey. Cone snails, turrids, murexes and most neogastropods are carnivores; some are generalists while others specialize in a single prey species. The most advanced cone snails have a radula modified into a hypodermic needle that is jabbed like a harpoon into fishes to deliver powerful toxins that paralyse them in seconds.

Cephalopods are the most advanced molluscs and have a distinctive head with well-developed eyes and nervous system. There are about nine hundred living species, and many thousands of fossil species; all live in the ocean and are predatory. Octopuses have eight arms, and most live on or near the bottom of the ocean, while squids and cuttlefish have ten appendages (eight arms and two long tentacles) and live in the mesopelagic zone of the water column. The most important fisheries of molluscs are those of squids and octopuses. Octopuses lack a shell, and squids have a reduced, internal shell called the pen, or gladius. The only living cephalopods with an external shell are species of the genus *Nautilus*.

Cephalopods have the most complex eyes among the invertebrates, and although they have evolved independently, they look similar in structure to vertebrate eyes. Giant squids have among the largest eyes of all animals: the largest specimen studied to date, with a body length of 8.5 metres (28 ft) had eyeballs that were 28 centimetres (11 in.) in diameter. The eyes of colossal squid, which reach 14 metres (46 ft) in length, are even larger.[7] In contrast, the blue whale, the largest known animal to have ever lived, has eyes about half of that size.

Squids are also the fastest of the invertebrates, while octopuses are known for their intelligence. Studies of animal behaviour in the field or in aquaria have shown that octopuses have sophisticated spatial navigation, and experiments have shown that they have memories and can learn tricks and play games.[8] Some scientists even believe that octopuses have consciousness.[9] Cephalopods have pigmented cells in their skin (chromatophores) that are controlled directly by nerve cells, allowing the animal to change colours in a fraction of a second; for example, a cuttlefish will use an amazing display of flashing colours during courtship to impress its mate. Octopuses are masters of disguise and can change not only the colour of their skin but its texture to mimic their surroundings.

The shell of the chambered nautilus, *Nautilus pompilius*, has been celebrated for its beautiful involute spiral, with stripped bands on the outside and a pearly interior. Although it is a logarithmic spiral, it is often erroneously cited as an example of the golden ratio. It is also an engineering marvel, composed of multiple chambers. The animal occupies the last and largest chamber. The series of earlier chambers, each hollow and sealed off from the others, is interconnected by a permeable tube, the siphuncle. The animal uses the siphuncle to fill this series of chambers with gas or water in order to control the animal's buoyancy, much like a submarine. The cuttlefish

*Cut nautilus shell (*Nautilus pompilius*) showing its pearly chambers.*

has an internal shell (called cuttlebone) with many chambers that hold gases and help to keep the animal afloat; the ram's horn squid, *Spirula spirula*, has an internal chambered shell that, like the *Nautilus'* shell, controls buoyancy.

The female of the paper nautilus, *Argonauta argo*, a small pelagic octopus, produces a paper-thin 'shell' in which it lays its eggs. This casing superficially resembles the shell of the *Nautilus*, but it is not chambered, or, in truth, a shell. The empty egg cases occasionally wash ashore by the hundreds, but they are very fragile, so perfect specimens are uncommon in collections.

Chitons or coat-of-mail shells (polyplacophorans) have a shell divided into eight articulated plates that move in relation to each other, allowing the chiton to roll up like a woodlouse when threatened. There are between 650 and 800 species of chiton, all of which live in the ocean. They are most diverse in temperate waters. Chitons are common on rocky shores in the intertidal zone, where they rasp algae from the rocks and use their muscular foot to crawl around and cling to rocks.

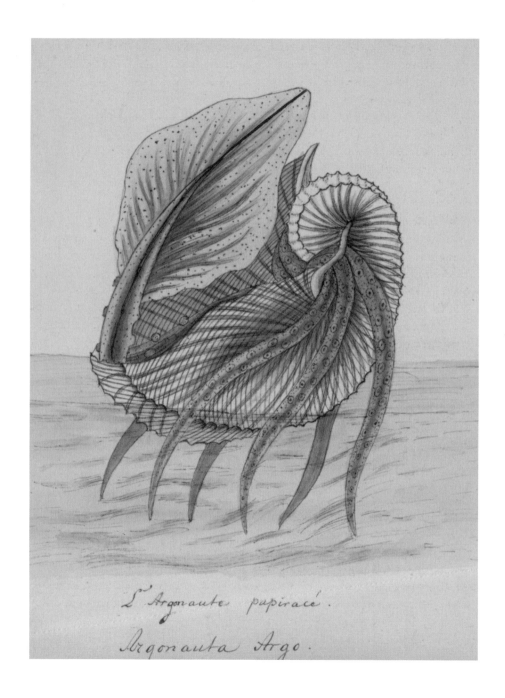

L'Argonaute papiracé.

Argonauta Argo.

Tusk shells (scaphopods) have a distinct elephant-tusk-shaped shell that is hollow and open at both ends. They live buried in sediment, with the narrow end sticking out of the surface and the wider aperture below. There are about nine hundred living species of scaphopods, with most species living in deep waters.

The other three classes of molluscs, Caudofoveata and Solenogastres (shell-less molluscs that resemble worms) and Monoplacophora (small limpet-like molluscs), are rare and live in deep water. Most species are minute; therefore, they are largely unknown to the public.[10]

Life Cycle and the Shell

The life cycle of a typical mollusc starts as a fertilized egg. Most marine molluscs have planktonic veliger larva, although some have direct development. The larva is carried by the ocean currents, and when it grows, it sinks to the bottom and metamorphoses into a juvenile. There is often a change in diet at metamorphosis that is reflected in shell shape and texture: if you look at a shell under a microscope, you can see where the larval shell ends and where the juvenile shell begins.

Once they reach maturity, marine molluscs typically mate or release gametes into the water, and the cycle starts again. The majority have distinct sexes and do not change them throughout their lives; in some species, however, individuals may start life as males and then change to females when they grow, or vice versa. Some molluscs, including land snails, are hermaphroditic, with both male and female reproductive organs present in the same individual throughout their life.

The shell starts to be deposited as a cap-like structure in the early larval stage and continues to grow larger by the addition of new material at its edge. The shell is secreted by the mantle, laid in thin increments that can be counted much like tree rings to estimate the age of the mollusc. The shell consists mostly of calcium carbonate ($CaCO_3$) arranged in two types of crystals: aragonite and calcite, which form prismatic or lamellar layers. Additionally, there is a protein matrix: many shells are covered externally by a 'skin' called periostracum, made of a tough but flexible protein (conchiolin) somewhat similar to that which forms our nails. The periostracum can be thin or thick, smooth or hairy, and often conceals the colour of the shell underneath. Inside many bivalve and gastropod shells there can be a pearly layer of nacre (mother-of-pearl). Most molluscs have the potential to

Paper nautilus, Argonauta argo, *a small pelagic octopus with arms extending outside of its egg case, 1800–1850, print engraving.*

secrete pearls, although relatively few species produce
iridescent pearls.

In gastropods the shell grows through the addition
of shell material to the aperture. Most gastropods have a
coiled shell with the aperture on the right-hand side when
the aperture is facing you and the apex is pointed upwards;
this is called a right-handed shell, because you can stick
the fingers of your right hand into the aperture and grab the
shell. A few gastropods, like the lightning whelk, *Sinistrofulgur
perversum*, have sinistral or left-handed shells. Chirality, or
orientation of coiling, is determined genetically; rarely, mutant
gastropods have a reversal in chirality, and normally right-handed
shells become left-handed (and vice versa).

Gastropods range in size from almost 1 metre (3 ft) long
(for example, the Australian trumpet, *Syrinx aruana*) to fractions of a
millimetre. Among the smallest gastropods is the minuscule ammonicera,
Ammonicera minortalis, which reaches only about 350 micrometres in diameter.

Opposite above: Lightning whelk, Sinistrofulgur perversum, *whose shell is left-handed, uncommon among gastropods. Opposite below: Egg mass of the lightning whelk. Above: Florida horse conch (*Triplofusus giganteus*), the largest gastropod in the Atlantic, and second-largest shelled gastropod in the world. Seen here is a side-by-side view of a juvenile and an adult, along with a bay scallop (*Argopecten irradians*) *on the far left.*

An intensive field sampling in the region of New Caledonia, which used different gears to sample all shell sizes, showed that the majority of the species had shells smaller than 8.5 millimetres (⅓ in.) in length.[11] These species are considered micromolluscs, often defined as having shells that do not grow larger than 1 centimetre (⅖ in.) long. They are diverse and abundant globally but are generally poorly studied throughout the world. They have been better studied in certain areas in Hawaii, for decades, and have been used as indicator species in a bio-monitoring programme to ensure that wastewater is properly treated before being pumped into the ocean. Among the community of micromolluscs in Hawaii there is a gastropod species that has an affinity for enriched sediments; a large increase in the frequency of that species, normally rare, could suggest eutrophication in the sediments near the outfall as a result of improper waste treatment.[12]

Bivalves range in size from minute to enormous; the largest is the giant clam, *Tridacna gigas*, which can grow to more than 1.2 metres (4 ft) in length and

Scanning electron microscope (SEM) images of Hawaiian micromolluscs less than 1 mm in diameter, with arrows indicating where the larval shell ends. The mollusc in the middle, Scaliola gracilis, *agglutinates fine sand grains to its shell.*

weigh up to 200 kilograms (450 lb). They are also known as giant killer clams owing to stories of swimmers getting their feet stuck between the thick valves of the giant bivalve and drowning in rising tides. While it is theoretically possible that fatal accidents such as these can happen, they are at most extremely rare since large clams close their valves slowly (this author has touched the velvety mantle of these rather gentle giants, and they only slightly closed their valves).

There are several species of very large squids known as giant squid and colossal squid; all are elusive and seldom seen alive. They captivate the public's interest and are the marine equivalent of dinosaurs in terms of popularity. They live in deep water and are mostly known from rare beach strandings in New Zealand, Japan and elsewhere, or from the stomachs of whales, their only predators. In recent years, submersibles and remotely operated vehicles (ROVs) have recorded a few glimpses of live giant squids. Their maximum size has been greatly exaggerated in the past, with reports of specimens as long as 41.5 metres (136 ft) being highly dubious. Currently, the maximum accepted size is about 14 metres (46 ft) in total length (measured from the end of the mantle to the tip of the tentacles), as found with *Architeuthis dux* and *Mesonychoteuthis hamiltoni*.[13] Contrary to popular belief and depictions, ranging from Pierre Denys de Montfort's *Histoire naturelle, générale et particulière des mollusques* (1801) to films such as *Pirates of the Caribbean: Dead Man's Chest* (2006), the mythological kraken of Norse folklore and sailors' stories is likely based on sightings of giant squids, and not of giant octopuses. The largest octopus species is the giant Pacific octopus, *Enteroctopus dofleini*, from the Eastern Pacific, which grows to a maximum of 6 metres (20 ft) in arm span.

Long Fossil History

Shells have been around for a very long time: the earliest fossils recognized as molluscs date from the early Cambrian (541 million years ago; earlier fossils that may be precursors to molluscs exist, but their identity has been debated). Because of their hard shell, molluscs have one of the best fossil records of all organisms. Several lineages have flourished and disappeared, leaving no recent representatives: the ammonoids, for example, were particularly diverse (some 10,000 species) between the Devonian and Cretaceous periods, but became extinct together with the dinosaurs at the end of the Cretaceous some 65 million years ago. Ammonite fossils can be quite abundant in certain regions such as the United States and Canada, and Madagascar.

The nautiloids were related to the ammonoids, but from a different branch in the molluscan phylogenetic tree. There were about 2,500 fossil nautiloid species; they disappeared hundreds of millions of years ago, and only a single branch survived

無名介

按アケマキノ一種ナラン

福田柳圃所藏

to present time in the Indo-Pacific Ocean. Currently only seven species of living nautiluses are recognized; the most common species is *Nautilus pompilius*. They are called 'living fossils' because their shells resemble those of their ancient ancestors.

The class Monoplacophora is one of the earliest branches of molluscs. It was believed to have become extinct in the mid-Devonian, until living specimens of *Neopilina galatheae* were collected off Suriname in 1952.[14] The discovery caused much excitement among both scientists and the public because the shells of the living specimens appeared to have been frozen in time, practically unchanged since their early appearance in the fossil record. Since the middle of the twentieth century about two dozen species of living monoplacophorans have been discovered from around the world.

Another example of living fossils are slit shells, a group of primitive gastropods in the superfamily Pleurotomarioidea. As the popular name suggests, the shell has a slit near the aperture. They first appeared in the Late Cambrian and were thought to have become extinct by the end of the Cretaceous, until the first living specimens known to Western scientists were dredged off Guadeloupe, in the Caribbean Sea, in 1858. The fishermen who collected it thought it was damaged and 'repaired' it by filing the shell lip before selling it.[15] Despite its lacking the natural slit, scientists used that specimen as the holotype of the new discovery; the shell is still at the Muséum national d'Histoire naturelle, Paris, today. As in the case of *Neopilina*, this was an important scientific discovery. However, unknown in Europe at the time, some eighty years earlier, the scholar and artist Kimura Kenkadō had published a hand-drawn book with an illustration of a recent species of slit shell from Japan.[16]

*Opposite: Woodblock print of a slit shell (*Mikadotrochus hirasei*) drawn by Kimura Kenkadō in* Kibai zufu *(Illustrations of Strange Shells, 1775).*

Above: Ecphora gardnerae *fossil, with a strong spiral sculpture that reinforced its shell against predators.*

CHAPTER TWO
TRIBAL SHELL USE

Humans have had a relationship with shells since the dawn of prehistory. In fact, new research shows that our ancestors, *Homo erectus*, created the first known doodle on a freshwater mussel shell about 540,000 years ago.[1] These early hominids used a shark's tooth to perforate a mussel shell and etch a zigzag pattern onto it in Trinil, on the island of Java, Indonesia. This finding pushes back the dates of the earliest known 'art' by more than 200,000 years. Whether this finding represents the beginning of art or is the result of a utilitarian process (say, to open the shell to eat

Opposite: Shell from Trinil, Jakarta, Indonesia, with zigzag drawing,
dated at 540,000 years old, the first known art; and detail, above.

the mollusc's flesh), it nonetheless suggests that shell collecting is probably one of the oldest hobbies. (And the fact that a humanoid had an interest in shells shows a high form of intelligence![2])

Molluscs that make shells were likely first used by humans as food. The earliest known evidence of shellfish consumption is from Terra Amata in Nice, France, and dates to about 300,000 years ago, before 'anatomically modern humans' (the term currently used to denote skeletons that resemble present-day humans) appeared.[3] Other early sites include several places in South Africa (with dates between 140,000 and 30,000 years ago), Spain (50,000–40,000 years ago), Australia and Papua New Guinea (both about 35,000 years ago) and Vietnam (33,000–11,000 years ago).

Shell Middens

Around the world, coastal communities, either inhabiting areas near rivers and lakes or, less commonly, inland, consumed large amounts of shellfish and discarded the shells in extensive heaps called shell middens or kitchen middens. Shell middens often are composed mostly of discarded shells, but there may also be fish bones or bones of other animals, and pieces of tools made from shells, bones and stones. The archaeological importance of these formations was first recognized in Denmark, where they are known as *køkkenmøddinger* (kitchen mounds), in the mid-nineteenth century. Shell middens may provide a wealth of information on native peoples, such as their diet, food-processing methods (for example, whether foods were baked, dried or steamed) and the seasonality of the shellfish consumed. There may also have been other uses of shell middens, such as burial sites or living areas.

Some of the oldest shell middens have been found in South Africa, from the Middle Stone Age (about 140,000 years ago). Maine in the United States is home to the largest-known shell midden in the country, the Whaleback Shell Midden, which was originally quite extensive: before part of it was processed into fertilizer in

Below: Shell midden on the island of Inishlacken near Roundstone, Galway, Ireland.
*Opposite: Queen conch (*Aliger gigas*) on the beach of St Croix, u.s. Virgin Islands.*

the nineteenth century, it contained an estimated 141,500 cubic metres (5 million cubic ft) of material, mostly oyster shells. Canada's Pacific coast features shell middens that are more than 1 kilometre (3,280 ft) long, and a midden in Namu, British Columbia – formed during more than 10,000 years of continuous occupation – is over 9 metres (30 ft) deep. In Brazil, *sambaquis* were once common along the coast; they were formed over 7,000 years until European colonization started in the sixteenth century. The composition of middens varies with location: some consist mostly of oyster shells or other abundant bivalves. On the Caribbean island of St Croix, U.S. Virgin Islands, the main species found in middens is the queen conch (*Aliger gigas*). Elsewhere, freshwater mussels or land snails may be the dominant species, in addition to the bones of fish or other animals, other leftover food and fragments of stone tools. Unfortunately, many of these sites worldwide have been destroyed due to mining for lime, coastal development, coastal erosion and, in certain places, land subsidence and sea levels rising.

Shell middens were originally considered to be evidence of the importance of shellfish to the diets of certain communities, given their extensive size. However, in the 1970s, adepts of the New Archaeology came to think that shellfish had less nutrition than the flesh of terrestrial animals, especially mammals.[4] Additionally, they postulated that the time invested in procuring shellfish for the protein yielded meant they would have been an inferior food choice, mostly consumed during famine events, when other more nutritious foods were not available. In recent decades, however, it has been recognized that seafood is indeed a valuable source of protein, amino acids and other nutrients; it also contains less saturated fat, meaning that it may be healthier than the meats of terrestrial animals. Furthermore, other arguments, such as the usually low risk to life when gathering shellfish and a lesser dependence on technology such as tools or weapons, weather, seasonality and expertise in comparison to hunting terrestrial game, have helped revive the credibility of shellfish as a bona fide food source in prehistoric times.

Delicious Molluscs

Today, molluscs continue to play an important role in human diets. Artisanal and commercial fisheries of many shellfish species are vital to coastal communities around the world. Alongside the traditional use of shellfish for subsistence, some molluscs have achieved the status of delicacies in modern societies and can fetch high prices (for example, abalone, opihi, giant clams and geoduck). Certain cultures also have deep cultural connections with shellfish (such as 'opihi in Hawaii).

Clams		
bittersweet clams	Glycymerididae	France, the Mediterranean, West Africa
donax clams	Donacidae	Australia, the Mediterranean, USA
geoducks	Hiatellidae	Alaska to Baja California, USA, Asia (imported)
giant clams	Cardiidae	Indo-Pacific, Japan (imported)
lucine clams	Lucinidae	Southeast USA to Brazil, Sweden
mactra and surf clams	Mactridae	Europe, Mediterranean, USA
mussels	Mytilidae	Americas, Europe, Oceania
oysters	Ostreidae	France, Japan, USA
pearl oysters	Margaritidae	Indian Ocean, Indo-Pacific
pen shells	Pinnidae	Black Sea, China, Europe
razor clams	Solenidae	Chile, China, Europe
scallops	Pectinidae	Americas, Europe, Indo-Pacific
soft-shell clams	Myidae	The Mediterranean, USA
tellin clams	Tellinidae	Europe, USA
venus clams	Veneridae	Europe, Southeast Asia, USA
wedge shells	Mesodesmatidae	Chile to Peru

Snails		
abalone	Haliotidae	Asia, Australia, New Zealand, USA
Babylon shells	Babyloniidae	Europe, Southeast Asia
escargot	land snail; Helicidae	Europe, especially France and the Mediterranean region, northern Africa
giant snails	land snail; Achatinellidae	West Africa
keyhole limpets	Fissurellidae	Chile, the Mediterranean
megasnails	land snail; Strophocheilidae	Brazil
murex shells	Muricidae	Indo-Pacific, the Mediterranean, South America
opihi, patella limpets	Patellidae	Europe, Hawaii, West Africa
periwinkles	Littorinidae	Americas, Europe
top shells	Trochidae	Americas, the Mediterranean
tritons	Ranellidae	Indo-Pacific, the Mediterranean,
true conchs	Strombidae	Caribbean, Indo-Pacific
tun shells	Tonnidae	Caribbean, Indo-Pacific
turban shells	Turbinidae	China, Japan
volutes	Volutidae	Argentina, Philippines, Senegal
whelks	Buccinidae	Europe, Southeast Asia

Cephalopods		
cuttlefish	Sepiidae	East Asia, the Mediterranean
octopus	Octopodidae	China, Europe, Japan, Mexico, USA
squid	Loliginidae	China, Japan, Peru, USA[5]

Shellfish gathering is a common activity throughout the world. Often done by women of all ages and by children, it is usually a leisurely activity that does not require special tools or skills. However, a few habitats, such as exposed rocky intertidal shores, pose hazards that do call for special skills and equipment. In Hawaii, *opihi* (limpets) are frequently gathered by men on steep rocky shores that are beaten by strong Pacific waves; fatal accidents are not uncommon. Perhaps the most extreme shellfish gathering occurs in Kangiqsujuaq, an Inuit village on the east coast of Ungava Bay in Quebec, Canada. Here, the villagers will crawl under ice sheets during the extreme low tides of the spring equinox to harvest protein-rich mussels; they have only about 30 minutes before the tides quickly return, and if they don't escape in time they will be crushed to death under the ice.[6] They risk their lives to add a little variety to their otherwise boring seal-meat diet, but the high stakes pay off when they find a bounty of delicious mussels.

Squid and octopuses are also molluscs, although the former have reduced and internal shells and the latter have lost their shells altogether. Large commercial fisheries of cephalopods exist in Asia, the Mediterranean and South America, and to a lesser degree in many other places. Artisanal octopus fisheries exist along the eastern Pacific coast from Mexico to Chile, and elsewhere. Land snails such as escargots (*Helix pomatia*, *Cornu aspersum* and other species) are gathered in large numbers in France, Spain and other areas where the snails are common. However, owing to commercial fishing pressure, habitat degradation, pollution and other environmental challenges, many shellfish populations are now greatly reduced and harvesting is no longer economically feasible. Some populations have had to be protected by laws to prevent their extinction, as the queen conch (*Aliger gigas*) has been throughout most of the Caribbean. Therefore, aquaculture is increasingly becoming more prevalent to supplement or replace certain fisheries.

Shell-Rich Culture: Papua New Guinea

Papua New Guinea is located on an island in the southwestern Pacific. Although it represents about only 0.1 per cent of the world's human population, more than 10 per cent of the world's languages are found there. The country is a hotspot of cultural and biological diversity, both on land and in the ocean. Being an island nation, its culture is intimately connected to shells, and even the areas far from the ocean, such as the highlands, value them highly. There are several Papuan examples throughout this book, such as the use of shells as currency, jewellery and musical instruments. In fact, a whole book could be dedicated to shell use in that country.

In many different places, trumpet or triton shells (*Charonia* spp.), conchs (*Aliger gigas*), helmet shells (*Cassis cornuta* or *C. rufa*) and other large shells are used to make trumpets. Typically, the apex of the shell is drilled or cut to make a blowhole, but sometimes the blowhole is drilled on the side of the shell. In most places around the Pacific, shell trumpets are made only with a carved shell but no mouthpiece; in New Zealand, however, the Maoris make *putatara*, a type of trumpet that uses a triton shell (*Charonia lampas rubicunda* or *Charonia tritonis*) and features a wooden mouthpiece. Shell trumpets are used to summon villagers on a small island or within small communities; to alert people of danger, such as an approaching storm; or to spread news to villagers – grave news, like the death of the chief, but also good news, such as the birth of a child.[7]

Shell trumpet blower, Milne Bay, Papua New Guinea.

In Palau and elsewhere in Melanesia, spider conchs (*Lambis lambis*), egg cowries (*Ovula ovum*) and other shells are hung in front of the tribal chief's house as a symbol of status. Other shells are worn as jewellery, such as large pearl oysters (*Pinctada margaritifera*) and several cowries, while the rare golden cowrie (*Callistocypraea aurantium*) is worn by the chief in the Marshall Islands. The ornaments can represent both one's wealth and one's power, or convey a person's rank in society; however, certain shells are restricted, to be worn only by the chief or his family.

Shell Tools

The Calusa were a tribe of Native American Indians that inhabited the Everglades and the southwestern Florida peninsula for about 3,500 years until their demise in the late 1700s. It is believed that their population reached as many as 50,000 people.

The word 'Calusa' means 'fierce people': the men were tall and well built, and ferocious warriors. They were the first Native Americans encountered by the Spaniards in Florida in the early 1500s, and early on they attacked the explorers. When not engaged in battles, the Calusa lived on the modern-day Gulf Coast and along the inner waterways of the Everglades. They had ready access to an abundance of shells, and many uses for them; because of this, they are known as the Shell Indians. The Bailey-Matthews National Shell Museum in Sanibel Island, Florida, features dioramas about the tribe,

Below: The Calusa, also known as the Shell Indians, lived on the West Florida coast and had many uses for shells, as depicted in this diorama at the Bailey-Matthews National Shell Museum, Sanibel, Florida. Opposite: Engraved shell gorget from Craig Mound, Spiro, Oklahoma, 900–1500 CE.

as well as several displays on the different tools and fishing gear made from shells. The Calusa also wore several shells as personal adornments, such as olive shells (*Oliva sayana*), which were perforated and strung as beads or used to decorate clothes. Large, nearly flat discs of whelk shells, called shell gorgets, were cut, decorated and hung around the neck, as part of the regalia worn by the chiefs (similar gorgets have been found in regions such as the Spiro Mound in eastern Oklahoma, which dates between the ninth and fifteenth centuries). Besides making tools from shells, the Calusa also engineered shell mounds around their villages into sturdy pilings built with large shells at the bottom and smaller shells on top. They dug canals to link waterways and created complex canal systems, which they travelled in dugout canoes. The canals were also used for fishing, the Calusa using several tools made from shells.

Shell tools made by Indigenous peoples around the world
Household dishes, cooking pots and utensils such as spatulas, cutlery and so on were made from large bivalve and gastropod shells
Storage containers, made from a number of large shells, were used to store medicine, ointments, pigments and so on
Fishing gear, including fish lures, octopus lures, hooks and sinkers, were made from many different bivalve and gastropod shells
Tweezers, tongs and claspers: many bivalves have nearly identical valves (equivalve) that may work well as tweezers
Weapons, made from sharpened parts of thick shells
Farming tools, shovels, plough blades and hoes, made from heavy parts of thick shells
Building tools designed to split and smooth various building and thatching materials such as palm fronds and bamboo canes
Adze, knife and axe blades were often made from sharpened shells of bivalves
Blades and scrappers, often made from bivalve shells, for example, ark shells, pearl oysters, but sometimes also with gastropods, for example, breadfruit scrapper made with cowrie shells
Drills, chisels and so on, made from pointy shells, for example auger shells
Oil lamps, in which oil is stored in a large shell and a wick floats on the surface, for example, giant clams (*Hippopus hippopus*) or spider shells (*Lambis* sp.)
Sanders were often made with textured bivalves, for example, file shells (*Lima* sp.)
Food pounders made from giant clams
Bailing buckets made from 'bailer' shells (*Melo* spp.) to remove water from canoes; some are still used by fishermen in Australia and the South Pacific

MATERIAL: COWRIE SHELL

USE: THE SKIN OF THE BREAFRUIT IS REMOVED BY SCRAPING THE FRUIT AGAINST THE SHARPEN EDGE OF THE HOLE

FEFEN ISLAND -

Shell Jewellery

Shells have been used as jewellery or adornment by humans and their ancestors since prehistoric times. The earliest known shell beads are from limestone caves in eastern Morocco: the Aterian archaeological sites date back some 110,000 years. Blombos Cave in South Africa also has some of the earliest shell beads, currently dated to 72,000 years ago. In both cases, small shells of the genus *Nassarius*, a group of marine and estuarine gastropods, were used. Recent analysis shows that the shells found in the caves were collected on the coast, then transported a long distance (20 kilometres/12 mi. in the case of Blombos Cave), perforated by a stone tool and strung as necklaces. Some of the beads from Morocco had red ochre residue. Microanalysis of the shells shows use-wear consistent with that of rubbing against thread, clothes and other beads, suggesting that they may have been passed from one generation to another. The distance from the coast, the irregular boreholes (different from the round boreholes made by predatory moonsnails), the presence of adult shells only and the pattern of use-wear confirm that the shells were indeed shell beads made by prehistoric humans.

Pacific Northwest Native Americans from about 1000 BCE to the present used *Dentalium* shells (known as 'tusk shells' or Dentalia), the shell of a scaphopod mollusc, as a form of currency and to decorate clothes and

Opposite: Pwil *breadfruit scrapers made with the sawn shells of tiger cowries (*Cypraea tigris*), Alele Museum, Marshall Islands.*
Above: Shell hammer/adze, Florida.
Left: Hawaiian tako *(octopus) lure made with a tiger cowrie shell, an oxen-bone fishhook and a stone.*

jewellery. However, tusk shells live in deep water and cannot be obtained without dredging or diving equipment. Ingeniously, the Native Americans developed a weighted device that was lowered from a boat by a rope into the muddy sediment in the deep ocean where tusk shells lived. The device had a central vertical shaft at the top where a rope was tied. The shaft continued downwards and was connected to a few planks of wood that were finely cut into a series of long rods. Those rods were the part of the device that penetrated the mud, entangling tusk shells between them. The device was weighted by rocks tied to a horizontal plate positioned about midway along the device's length. The species *Antalis pretiosa* was the most sought after because of its large size, up to about 55 millimetres (2.2 in.) in length. It would be collected off Vancouver Island, where it was most common, but it was never found in large numbers, and therefore became seen as rare and valuable. Several Native American tribes obtained tusk shells through trade and likewise used them as jewellery and to adorn clothes. They were fashioned into bands or belts, with two or more parallel rows of shells arranged horizontally or diagonally. Later, shells were replaced by elongated beads made from bones that resembled tusk shells; bone beads then became more common in large breastplates made by different Native American tribes after the 1880s.

Another use of shells by Native Americans was to fashion beads called *wampum* from shells. The most valuable *wampum* was made from the purple parts of the clam *Mercenaria mercenaria*, native to the eastern United States. The clam has a large, thick shell, but only a small portion of the shell, near where the muscles that close the valves attach, is purple. Since the purple part was scarce, it was more valuable than the rest of the shell, which was white. Beads were carved as short or long tubes and drilled in the centre to be strung into clothing, jewellery and, most notably, *wampum* belts. Artists used the purple and white beads to make designs, and one of them, the *kas-wen-tha*, or two-row *wampum* belt, has a special importance and symbolism. It represents a

Opposite: Tlakluit bride from the Wishram tribe, Columbia River, Oregon, wearing braids, beaded headdress with Chinese coins, dentalium shell earrings, beaded buckskin dress, c. 1910, photograph by Edward S. Curtis. Above: Mercenaria mercenaria *shell.*

diplomatic agreement between the Iroquois (Haudenosaunee) tribe and European explorers and settlers, and its message welcomed the white men to the tribe's land: 'We will not be like father and son, but like brothers.' The two purple rows symbolize vessels (the culture, laws and customs of each nation) travelling down the same river together, one vessel being the Original People and the second, the Europeans. The vessels would travel the river side-by-side, without interfering with each other.

Entrepreneurial settlers used other thick shells to produce *wampum*. One of the shells was the queen conch (*Aliger gigas*) from the western Atlantic. The thick lips of queen conchs were broken and then polished and carved into long shell beads called hair pipes. The difficult part was to drill a hole through the long bead. The technology available at the time only allowed about half of the length of a bead to be drilled, so the beads had to be drilled from both ends in a straight line and ideally meet in the middle. Documents from the 1830s suggest that in some instances beads were not completely bored through and would be rejected by the buyers. Hair pipes were traded with Native Americans, sometimes far from the coast, who used the long beads to decorate breastplates and other clothing. In central and southern California shells of the small olive snail (*Callianax biplicata*) were used for decoration over a period of 9,000 years. Between the sixteenth and nineteenth centuries olive shells were also used as money on the east coast (some tribes used other species, such as the marine bivalve *Saxidomus*).

Shell jewellery can be found in almost every culture throughout space and time. One would expect that communities near shores would have easier access to shells and, therefore, use more shells than those communities further away; that seems to be the trend, but there are notable exceptions. For example, the highlands of Papua New Guinea highly prize shells, especially the more colourful seashells that come from faraway shores. Native Americans throughout the United States and Canada, even tribes far inland, obtained shells and *wampum* though trade. In high-altitude Nepal and Tibet some shells, like the sacred chank, and shell beads are an important part of the culture and religion. Throughout Africa, shells are present in many aspects of life, including food, religion, currency and the arts. Shell jewellery is an intersection of some of these cultural aspects. Among the different shells, cowries are the most prevalent type used in African jewellery: they can be found strung as necklaces, bracelets or anklets, woven into fabrics and belts, braided into hair, fashioned into earrings and so on.

Shell gorget, tools, wampum *and other shell objects made by Native American peoples of the southern* USA, *illustrations from Charles C. Jones Jr,* Antiquities of the Southern Indians . . . *(1873).*

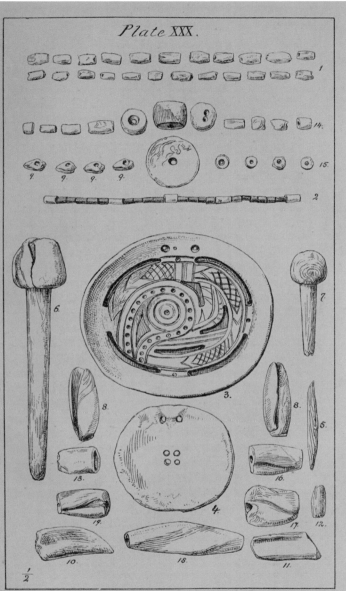

Plate XXX.

Cowries symbolize many things, such as fertility or prosperity, and are an important part of African culture.

Even today, shells continue to be used by both Indigenous and non-Indigenous peoples for different purposes, but they feature most prominently as jewellery. Shell jewellery can be used to adorn both the living and the dead; it can be bland or colourful, rough or polished, whole or cut. Pieces can be used to tell one's rank in society, or to show whether one is available and/or interested in the other sex. They can be used as protection from evil spirits or to bring good luck; to aid fertility or to wish for an easy delivery during childbirth; to display one's mood; or simply worn because they are beautiful objects of art. Shell jewellery often includes mother-of-pearl and pearls.

*Opposite: Hat (*kalyeem*), Kuba people, Central Africa, early 1900s, raffia, glass beads, cowrie shells, cloth (including wool), thread, copper alloy.*
*Below: Royal belt (*yet*), Kuba people, Central Africa, mid-20th century, leather, cotton, glass beads, seashells, cowrie shells, brass, twine, pigments.*

Mother-of-pearl (or MOP) is a naturally occurring nacreous layer found in many shells (although not all shells have nacreous layers). Pearls are the only gems that are made by living organisms. Modern shell jewellery can be found at local handcraft markets and department stores, or from all over the world via the Internet. (Whenever possible, however, it is best to try to support local communities and native artists.)

Shells at Burials

Alongside their use as jewellery and decoration for the living, shells were also used to decorate the dead. One early burial with shells from Qafzeh, Israel, is about 92,000 years old.[8] The seashells at this burial site came from 35 kilometres (22 mi.) away. While early burials with shells are rare, they are very common on Upper Palaeolithic burial sites in different countries. A report of a more recent burial of an infant, found at a site from the Upper Palaeolithic Magdalenien culture in France 10,200 years ago, reports that at least 900 *Dentalium*, 160 *Neritina*, 36 *Turritella* and 20 *Cyclops* shells were found at the head, elbows, wrists, knees and ankles of the child. The burial site and the large number of shell pendants suggest that the infant had a high social standing.[9] Among the many different uses of cowrie shells was the practice in ancient Egypt of placing cowrie shells in the eye sockets of the deceased to help them see the path in their journey to the next world.[10]

In a more recent burial from the late Mesolithic and early Neolithic of Téviec – an island on the coast of Brittany, France – ten multiple graves with the remains of 23 people were discovered in the 1920s. Current dating methods place them between 6,740 and 5,680 years BP. The people were buried in an extensive shell midden, which helped to preserve the human remains from dissolution in the acidic soil. Besides human skeletons and shellfish, there were remains of fish, birds and terrestrial mammals, including antlers, as well as many flint tools. The skeletons of two women, the 'Ladies of Téviec', who died violent deaths, were decorated with shell jewellery attached to their legs and necks, as well as other shells found in the grave, covered with a roof of red deer antlers. The grave was excavated and moved to the Muséum de Toulouse, France.

Tyrian Purple and Other Molluscan Dyes

Tyrian purple, named after the city of Tyre in Lebanon, is a natural dye from the secretions of the hypobranchial gland of gastropods of the family Muricidae. The dye was used by the Phoenicians to produce prized cloths with a blue to reddish-purple tint that was lightfast (that is, would not fade when exposed to sunlight). Because of the great expense needed to produce it, purple-dyed wools were worth their weight

Skeletons of two women buried with shell jewellry, site of Téviec in Brittany, 6740–5680 BCE (Mesolithic).

in gold, and the dye was reserved for royalty. Purple then became recognized as the royal colour, hence the ceremonial robes of Byzantine emperors, European royalty and Roman Catholic bishops. In the Mediterranean Sea, the most important muricid species used in the production of purple dyes were *Bolinus brandaris* and *Hexaplex trunculus*, but at least two others were used at times. Massive numbers of snails had to be procured and sacrificed to produce the dye; the shells were crushed and sea salt was added to the macerated mass, which was then kept in large stone or iron vats for three days and cooked for up to ten days. After an elaborate and foul-smelling process, cloths were dipped into a reddish-tinged solution; when dried and exposed to the sun, the cloths turned a deep magenta. Today billions of crushed shells can be found in mounds near the ancient murex factories. The process was labour intensive and kept a secret, even though Pliny the Elder described the process in his *Natural History* (77–9 CE). The origin of the purple dye industry is disputed; some authors point to Crete before 1750 BCE, while others propose a Minoan origin between 2000 and 1600 BCE.[11] The production of Tyrian purple ended with the Siege of Constantinople (present-day Istanbul, Turkey) in 1204 CE, during the Fourth Crusade. With the dearth of Tyrian

Byzantine emperor Justinian clad in Tyrian purple and pearls, 546–56 CE, mosaic, Basilica di San Vitale, Ravenna.

purple in Europe, vermillion, a brilliant red pigment made from an insect (*Kermes vermilio*) was adopted instead.

In western Mexico a similar process using another muricid snail, el caracol púrpura (*Plicopurpura columellaris*), has been in production since at least pre-Columbian times. The process is similar to the one employed in the Mediterranean, but instead of crushing the shell, Mixtec people gently 'milk' the snail over spools of cotton threads that eventually become dyed. After milking the snail, it is returned to its rocky habitat.

The milking is a slow process that requires training and patience. Promoters of this practice claim it is a sustainable activity, but the numbers of the snails are decreasing due to poachers and increased tourism. It is estimated that about 1,000 snails need to be milked in order to dye only 113 grams (4 oz) of thread (in contrast, it is said that 12,000 *Bolinus brandaris* were needed to produce only 1.4 grams (0.05 oz) of dye, not even enough to dye one whole garment). The resulting purple cloths are similar to those produced in antiquity; the more times the cloth is treated with the snail secretion the darker the cloth. Like the Europeans of antiquity, Mexicans also value the colour purple above all others, but tourists do not appreciate it – or at least are unwilling to pay a lot of money for a naturally produced purple cloth. The practice, therefore, is a vanishing art.

There are other species that are able to produce dyes which could be used to colour cloth. In Caribbean Mexico, the related species *Plicopurpura patula* produces a similar purple dye. The dogwhelk (*Nucella lapillus*), from the North Atlantic, can produce reddish-purple and violet dyes; and several species of epitoniid snails (wentletraps) can produce purple stains, as can some species of *Aplysia* when distressed, although they have not been investigated and exploited commercially for the production of dyes.

Other Shell Uses

Shells have been used for many purposes and products over the centuries.[12] Shells of abalones, pearl oysters, freshwater clams and *Trochus* species, as well as other shells with nacreous layers, were used to produce mother-of-pearl buttons, until plastic buttons replaced them in the 1940s.

The translucent and nearly flat shells of the windowpane oyster or capiz shell (*Placuna placenta*) have been used for thousands of years as a substitute for glass in windowpanes, lampshades and so on in China, Japan, the Philippines and other Asian nations. Likewise, *Anomia* spp. are clams that have thin, semitranslucent valves; the animal is edible, and the valves are used to make strands of shells that act as wind chimes, hence the popular name 'jingle shells'.

Cosmetic body art is one of the earliest demonstrations of human culture, potentially originating more than 100,000 years ago. Shells were used as containers to store ochre and other pigments used in early make-up, such as those found at archaeological sites at Ur (2600 BCE). Several games from antiquity also used shells as game pieces: the popular game Go from China, for example, or the Royal Game of Ur (2600 BCE), a game with a board made with inlaid shell plaques.

CHAPTER THREE
SHELLS AND RELIGION

S ince the dawn of human civilization, shells have been a part of religion. Different cultures around the world have attributed special powers or symbolism to shells, and many shells have been used in connection with religious ceremonies. In this chapter, we shall explore some of these rituals. As mentioned earlier, shells had many different tribal uses, among them personal

Opposite: Hupa shaman, Pacific Northwest, with necklaces of tusk shells and strands with bivalves on her head, c. 1923, photograph by Edward S. Curtis. Below: Incised conch shell trumpet, Chavín culture, Peru, 400–200 BCE.

Left: Relief of feathered conch shell trumpets known as caracoles emplumados, *Teotihuacan. Opposite: Statue of the god Vishnu holding a sinistral sacred chank shell in one of its hands, Nepalese, c. late 15th–16th century, mercury-gilded copper alloy with semi-precious stones, rock crystal and glass.*

decoration for the living and the dead. In some Palaeolithic burials in Europe, the Middle East, Africa and elsewhere, shells were strung into necklaces, bracelets, anklets and so on. Anthropologists believe that shells were used as amulets to protect the dead and to help them transition to the afterlife. In Qafzeh Cave in Israel, Blombos Cave in South Africa and at other sites,[1] different shells were found holding deposits of red ochre, a common natural pigment that contains iron oxide. Ochre comes in various colours and shades, and red ochre was used widely as body paint and in cave art because it symbolized blood and rebirth.

Likewise, as we have seen, large gastropod shells have been used as trumpets throughout the Indo-Pacific, Tibet, India and elsewhere, for different purposes including for religious ceremonies. In the Chavín culture, a pre-Columbian culture of coastal Peru, carved and incised conch shells were used as trumpets, probably in ceremonial rituals. In Mexico, Teotihuacan was the largest pre-Columbian city (near present-day Mexico City). Within the extensive complex that includes the Pyramid of the Sun and Pyramid of the Moon, archaeological excavations have revealed many buildings decorated with large conch-shell trumpets with delicate feathers (known as *caracoles emplumados*), alongside other shell decorations carved in stone or painted on murals.

The Sacred Chank

The sacred chank (*Turbinella pyrum*, family Turbinellidae) is a shell that is venerated by followers of two religions: Hinduism and Buddhism. It is a massive shell, reaching 25 centimetres (10 in.) in length and over 1.5 kilograms (3.3 lb) in weight. It occurs commonly in southern India and Sri Lanka, where there are large fisheries for this shell, and millions are fished every year. The popular name is derived from the Sanskrit word *shankha*, which means 'divine conch'.

In Hindu mythology, the preserver god Vishnu (or his incarnation as the four-armed god Krishna) is often represented in paintings and statues holding a

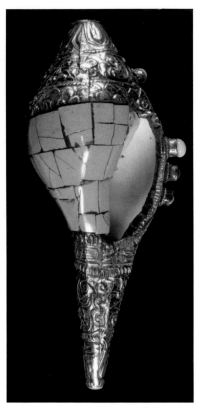

sinistral sacred chank.[2] As early as 3,000 years ago this shell was already used as an ornamental or religious object in northwestern India.[3] Shells were cut and used as good luck amulets, bangles and war trumpets. According to the *Bhagavad Gita*, one of the most sacred Hindu scriptures, each warrior had their own personal conch, distinguished by its colour, size and sound. The most important conch of all was Krishna's *Panchajanya*, which was small and slender, and produced a sweet sound. According to one version of the legend, the demon Panchajanya lived in a shell at the bottom of the sea, and Krishna acquired his conch after slaying the demon. In other tales, Krishna blew his chank trumpet to greet the king, or used it to announce a battle. The sound of the sacred chank symbolizes the sacred sound *Om*.

Most gastropod shells are dextral, or right-handed; left-handed or sinistral shells coil in the opposite direction. Mutations may cause a reversal of chirality, which is rare in some species. Because of its economic importance, accurate records are kept on the Sri Lankan chank fishery, so it is known that only one in about 600,000 chanks is left-handed, making them very rare and desirable (by contrast, 1 in every 440 shells of *Prunum apicinum*, family Marginellidae, is sinistral).[4] Despite their rarity, there are probably several hundred sinistral chanks in temples and private collections in India; however, there are apparently only three in American shell collections.[5] Wealthy people in India are willing to pay small fortunes for a sinistral chank, known as *Dakshinavarti shankh* or *valampuri*. One such shell fetched U.S.$45,000, making it one of the most expensive shells in the world.[6] Unfortunately, con artists fabricate sinistral chanks, for example by using shells of the lightning whelk, the official state shell of Texas, an unrelated and common sinistral species from the Gulf of Mexico and Atlantic Ocean. A fake shell can be easily disguised when it is covered in semiprecious stones, silver and other ornaments. Shells not covered in metals can be X-rayed to reveal

internal structures (columellar folds that continue into the shell) that distinguish the sacred chank from the lightning whelk; however, the lightning whelk is surprisingly popular in India among the masses that cannot afford the much rarer sinistral chank shell.

Chanks are also important to Buddhists, who associate them with the deities Sagramati and Gandhahasti.[7] They are often heavily decorated and mounted in silver, bronze or tin, and used as trumpets in religious ceremonies.

*Right: Conch shell with carved decoration used as a ritual water vessel for worshiping Vishnu, Indian, c. 11th century. The intricate carving does not go through the shell. Opposite: Sacred chank shell (*Turbinella pyrum*) alongside one from Tibet that is decorated.*

In Buddhism, chanks represent one of the eight auspicious symbols, the *Ashtamangala*, but sinistral chanks are considered particularly special and said to be worth their weight in gold.

In Hindu art, chanks are also carved into intricate designs, often depicting Vishnu. They are a symbol of water and are typically associated with female fertility and serpents. Some carved chanks were used as containers to transport water, while others were used as trumpets. Some authors believe that Hindu influence reached the Mesoamerican Aztec culture, which also revered large shells, including one related to the sacred chank, the West Indian chank (*Turbinella angulata*), which can grow to almost 50 centimetres (20 in.) in length.[8] However, further research is needed to confirm or refute this theory.

St James Scallop

The shell known as St James scallop (*Pecten jacobaeus*, family Pectinidae) is another species that is intimately connected to religion, in this case Christianity in Europe. Since medieval times, this scallop has been chosen to represent pilgrims on the Camino de Santiago, or the Way of St James, one of the major Christian pilgrimages. The pilgrimage dates from the ninth century, when the first pilgrims started to visit the shrine dedicated to St James at Santiago de Compostela in northwestern Spain, where his tomb is located. At the time, it was part of the Kingdom of Asturias and Galicia, and pilgrims began bringing home a common shell found on the Galicia coast as proof of completion of the pilgrimage. By the twelfth century, large numbers of pilgrims were visiting the shrine, and the pilgrimage became an organized affair. The scallop was adopted as a badge of the pilgrims for several reasons, including recognition by other pilgrims; to receive service at hospitals and hostels; and as a souvenir. Typically, pilgrims attached a scallop shell to their clothes, hat or walking stick, and the practice continues today: scallop shells can be collected at the seashore in Galicia or bought at souvenir shops along the way.

The scallop shell has many symbolisms, including the numerous routes of the pilgrimage being mirrored in the converging lines of the shell. Other, more practical reasons for the adoption of the scallop may have included it being lightweight and easy to carry, or to serve as a bowl or a cup to present in churches and other establishments along the way. (Typically, one shell-full portion of food or drink was donated to pilgrims that offered up a scallop shell.)

Juan de Juanes, St James the Pilgrim, 1560–70, oil on board.

The Way of St James has many routes of varying lengths (for example, the nearly 800 kilometres (500 mi.) route from Roncesvalles to Santiago de Compostela through León, or the Portuguese way, about 610 kilometres (380 mi.) from Lisbon) and several originating countries, but the majority end in Santiago de Compostela. A few brave pilgrims (less than 5 per cent), however, continue beyond Santiago de Compostela to Finisterra (or Fisterra in Galician, meaning 'The End of the World' in Latin), some 88 kilometres (55 mi.) further to the west, ending at the Atlantic Ocean. Among the popular routes, several start in France, Spain, Portugal, Italy, Germany and beyond. Today, local governments try to revive the traditions and encourage tourism. Many of the routes are prominently marked with signs of scallop shells or a stylized yellow logo against a blue background. The number of pilgrims completing the trip varied from few in the mid-eleventh century to many in the centuries that followed, but it declined after the Black Death. In the early 1980s, only a few dozen people completed the pilgrimage, but the

Left: Pulpit in the shape of a giant mahogany sculpture of the pilgrim's scallop at St Louis Cathedral, New Orleans. Above: Sign for Camino de Santiago de Compostela (Way of St James), Casco Viejo, Bilbao, with a stylized scallop shell. Opposite: Coat of arms of Pope Emeritus Benedict XVI with a pilgrim's scallop, Ottobeuren Abbey.

number has increased steadily since then to a peak of 272,703 people in 2010. Reasons for attempting to complete the pilgrimage included plenary indulgence, contemplation and soul searching, and pilgrims who walk at least 100 kilometres (62 mi.) receive a certificate of accomplishment, known as a *compostela*, issued by the church in Santiago. Scallops also appear in several cultural and artistic manifestations, such as the fine arts, heraldry and more. At least a few shells have been used in heraldry, but the scallop is the most common. Some Western examples of coats of arms with scallops include that of Pope Emeritus Benedict XVI, Princess Diana (three scallops) and Prince William (a single small scallop), but scallops feature in coats of arms in different countries and regions across the world.

Giant Clams

The largest shelled living mollusc, the giant clam (*Tridacna gigas*, family Cardiidae), can reach nearly 1.2 metres (4 ft) in length. The shell is massive and can weigh more than 200 kilograms (440 lb). The giant clam lives in the tropical eastern Indian Ocean and the South Pacific, in shallow water, and both its meat and its shell are popular for food and decoration, respectively, in local communities. Because of this, however, it is now listed as vulnerable by the International Union for the Conservation of Nature (IUCN) and included in the Convention on International Trade in Endangered Species of Wild Fauna and Flora (CITES) list, an international treaty. Thus it is now illegal to export the mollusc to other countries. (Farm-raised specimens with proper documentation can be exported, but they rarely reach giant sizes.)

Before international trade of giant clams was restricted, some large shells made their way to churches in Europe, where they were used as large baptismal fonts or basins. The French word for giant clam is *bénitier*, which is also the word for holy water font. One of the finest examples of a *bénitier* is the font at the Church of Saint-Sulpice in Paris. The huge shell was donated by the Venetian Republic to King Francis I of France

Opposite: Baptismal font shaped as a giant clam, c. 1745, sculptor Jean-Baptiste Pigalle, Church of Saint-Sulpice, Paris. The marble pedestal is carved with several marine creatures, such as the octopus below the shell.
Right: Queen conch set in stone for use as a holy water font, Convent of the Immaculate Conception (Las Monjas), San Miguel de Allende, Guanajuato.

in the early sixteenth century. King Louis xv commissioned the pillar, expertly crafted by the sculptor Jean-Baptiste Pigalle, who created an interesting marble piece decorated with several marine organisms. The basin is made from a real giant-clam valve, with its rim covered in brass to protect it.

Some churches have holy water fonts shaped like a giant clam (for example, the Church of Santa Maria degli Angeli in Rome) or a scallop (the Church of Saints Cajetan and Maximillian in Salzburg, Austria). Smaller fonts, called stoup, are often shaped like scallops. In the Convent of the Immaculate Conception (better known as Las Monjas Church) in San Miguel de Allende, Mexico, an unusual holy water font has been made from the shell of a queen conch (*Aliger gigas*, family Strombidae) that has been cut open and embedded into the stone wall.

Cowries

The shells of cowries, gastropods in the family Cypraeidae, have long been associated with female genitalia and fertility. Hence, cowries are sewn into the hems of women's clothes in Africa and India as an amulet for fertility and love. Cowries also feature in small fetishes (objects that have special powers, or that a god lives in), thus becoming part of religious rituals, and some ceremonial garbs are also ornamented with the shells.

Cowries are used in shell divination as part of religious rituals in several African religions throughout West Africa, Brazil and the Caribbean. The most common shell divination set-up in Brazil, called *jogo de búzios* in Portuguese (literally, 'game of shells'), or *Merindinlogun* (from the Yoruba language, meaning 'sixteen'), consists of sixteen cowrie shells that are thrown on a mat, table or floor, within a circle. Often the shells have the dorsal part ground up so that there is a large opening ('open shell') in addition to the narrow opening of the shell ('closed shell'); note, however, that some people consider the opposite, that is, the natural narrow opening as open and the other side closed. Before throwing the shells, the priest (or priestess; *Babalorixá* in Brazil) invokes the gods (*orishas*), asks questions and throws the shells. The gods intervene, influencing how the shells are arranged to answer the question; the priest then interprets the combination of open and closed shells and provides the answer. The number of shells varies from 4 to 21. Besides cowrie shells, other shells, rocks, bones, peppers or other objects can be used, but cowries are the most common. Variations of this shell divination ritual are used in the religions Yoruba, Candomblé, Santería, Umbanda and several others. *Jogo de búzios* is extremely popular in Brazil and parts of the Caribbean where there is strong African influence. Besides cowries, many other shells are also used in religious rituals in African religions.

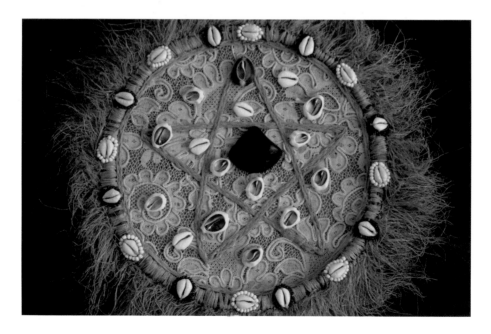

Above: Jogo de búzios *(*Merindinlogun*), Salvador, Bahia, Brazil.*
Opposite: Queen conch shells used in a Santería altar (Yoruba religion), Trinidad, Cuba.

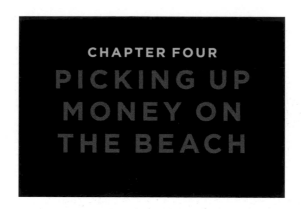

CHAPTER FOUR

PICKING UP
MONEY ON
THE BEACH

I magine going to the beach to pick up money when you need it. We have already discussed how some of the most ancient forms of currency were shells, but most people may think it surely happened long ago and only in a few places. They probably would be surprised to learn that in some places, shells continue to be used as currency.

Tabu and Other Shell Money in Papua New Guinea

One place where shell currencies continue to be used is the Tolai region in East New Britain, Papua New Guinea, where small shells of *Nassarius camelus* and *N. fraudulentus* (family Nassariidae) are used as official currency, Tabu (or Tambu), in parallel to the official paper currency, Kina. Tabu has been used for generations as traditional shell money, and in the late 1990s it was established as an official currency in the East New Britain region. It has since been regulated so its value is maintained as stable in relation to Kina (although the Kina has dropped in value). A large wheel, or

loloi, may have about 600 fathoms (1.1 km) of Tabu (somewhere between 180,000 to 240,000 shells) and be worth about k2,000 (or about U.S.$600). In 1998 it was estimated that there

Left: Close-up of the currency, showing
how shells are strung together.
Opposite: Boy from the Tolai tribe,
Karavia liu village, Papua New Guinea, with
the Tabu currency that is still in use today.

Opposite: Shells are still used as a symbol of status in the Goroka tribe, Papua New Guinea. The large discs at the bottom are a currency, Kina, *made from large shells of the pearl oyster,* Pinctada margaritifera. *Right:* Kueruek, *a ring made from a giant clam shell is used as a currency by the Abelam people, Papua New Guinea. It is shaped to resemble a boar's tusk, which is another form of currency.*

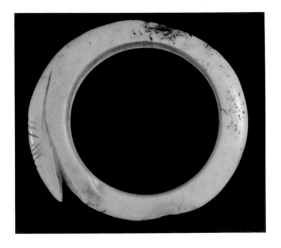

were some 4 million fathoms of shell money in the Tolai region, although only about one-quarter was in circulation, while the rest was stored in rolls or wheels by the elder clansmen.[1]

Shells are collected in the sea, cleaned, perforated and strung with varying lengths of rattan. The value of Tabu depends on the length (measured in fathoms or *pokono*), the number of shells per length and the quality of the shells. White or yellow shells are the most valuable, followed by brown; black shells are not used. Everyone in the village participates in the production of Tabu: children prepare the shells; women string them; and men wind Tabu onto wheels. Once Tabu is rolled onto wheels and bundled (*loloi*), it is stored and reserved for special occasions, such as a wedding or funeral. At the funeral of a clan leader, it is customary to cut open a *loloi* and distribute it to those attending based on their social status. Short lengths of Tabu are used daily to buy goods or pay fines, taxes and so on. Short lengths of Tabu can be attached together or restrung to make longer lengths.[2]

Several different types of shell currency have been used in Papua New Guinea, some to this day. Among them were large pieces of pearl oyster shells, called *Kina* (which gave a name to the paper currency). The value of *Kina* shell pieces was proportional to their length. Today, *Kina* shells are used more as a symbol of status. Another shell currency, *Kueruek*, used at least until the early 1900s in the Abelan region, was made from giant clam shells that would be polished and shaped like a ring or bangle large enough to be worn on the wrist. The ring would also have had a carved pointed part that resembles a boar's tusk (another form of currency).

The Abelam people made larger, polished discs from giant clams that were used as another shell currency. Elsewhere, in the Milne Bay region, *Bagi* necklaces were made from red shells and used for trading, as were *Mwali* (white shell armbands). In the New Britain province of Papua New Guinea and the neighbouring Solomon Islands, *Tafuliae* shell money is made from small discs or beads cut from reddish-brown shells and strung into strands. Different cultural groups have different names, shells and lengths for the shell strings. They are still used in many places, where they are required for ceremonial payments. Other shell money used in Papua New Guinea includes the *Kakal* and *Mis*, used by the Mengens (Pomio) and Baining people, respectively.

Money Cowries

The best-known shell currency was the money cowrie (*Monetaria moneta*, family Cypraeidae), a small marine gastropod with one of the widest distributions among cowries, ranging from tropical East Africa to Cocos Island in the Eastern Pacific. Its shell is typically about 25 millimetres (1 in.) in length, but it can range from about half of that size to nearly double. However, most shells used in the shell trade were similar in size, and larger money cowrie shells were less acceptable as currency because of the increased weight to transport.[3] The shell is glossy, yellow on top and white on the sides and base (ventral side), with a few dorsal bumps, and a long, narrow aperture on the ventral side. It was first used as currency in northwestern China about 4,000 to 5,000 years ago, brought through by traders from coastal areas, but its use expanded and peaked between 4,000 and 3,000 years ago, during the Shang and Western Zhou dynasties.[4] Around that time, cowrie imitations were made of stone, copper, bone, clay, and even silver and gold. Once metal coins became

Opposite: Sepik man statue sitting on several discs made from giant clam shells, Tridacna *sp., Abelam tribe, Papua New Guinea. Below left: String of money cowries, c. 1,000 BCE, Zhou Dynasty, China. Below right: The best-known shell currency, the money cowrie (*Monetaria moneta*), was once the most widely used currency in the world.*

more common in China, around 345 BCE, cowries were discontinued as currency. However, their use resumed some three hundred years later because coins were being counterfeited. A testament to the influence of cowries can be seen in the Chinese language: the character for cowrie (pronounced *bei* or *bui*) is a radical that is part of many words, including 'treasure', 'riches', 'high value', 'money', 'coin' and 'purchase'.

Centuries later, money cowries (and to a lesser degree its relative, the ring cowrie, *Monetaria annulus*) became the most widely used form of currency among trading nations of the Old World, as well as island nations in the Indo-Pacific. The Maldives, in the Indian Ocean, became the most important source of money cowries, and Bengal its distribution centre. Arab traders brought shells by caravans to inland Africa, and eventually they reached West Africa, where these cowries do not occur.

Left: Ancient copies of the money cowrie in copper (the three cowries on the left; largest about 30 mm in length), deer bone (right side, bottom) and stone (right side, top) from China, which were made where the shell was scarce. Today these are far more valuable than the real shell.
Right: String of money cowrie shells, 19th century, Maldives.

There, they were worth much more than in the Maldives, and the trade was highly lucrative, with profits reaching as much as 500 per cent.[5]

Advantages of the use of cowries as currency include the shell being lightweight and very durable, easy to carry and count, and hard to counterfeit. They could also be obtained in large quantities in certain areas, such as the Maldives, and could therefore be used as ship ballast. (Even bilge water would not damage them: a simple wash and the shells would be clean and glossy again.) Cowries are beautiful, and a pleasing size and weight to hold in your hand; however, unlike metal coins, the same shell cannot represent higher values. Therefore, in order to make high payments, large numbers of shells are required, which can be challenging to carry. Although there are over two hundred species of cowrie, other cowrie species (bar the ring cowrie) never really caught on as currency. A few large cowries were used in some places to represent larger values, but only locally. There is evidence that suggests the money cowries from the Maldives were significantly smaller than elsewhere, but some authors disagree. In general, it is believed that most cowries in the shell trade were of similar size and weight, which then made it easy to estimate the value by weighing. For example, in Bengalese markets, large baskets holding about 12,000 shells were used to make large payments.

Money cowries are linked to a dreadful part of history. Between the sixteenth and nineteenth centuries, Europeans became involved in the slave trade in Africa; the Portuguese were the first among the traders, followed by the British and Dutch, and later by the German and French, among others. Several African nations were also directly involved in the slave trade. Initially, the shell money was cheap, so the trading ships loaded millions of them as ballast in India and Sri Lanka, along with goods like silk, spices and tea. Then they would travel back to Europe, where the shells were auctioned and loaded back onto other ships, which then sailed down to West Africa, where the shells were used to buy slaves to work in plantations in the Caribbean. The price for an adult male slave was about 10,000 cowrie shells in the 1680s, but within less than a century, the price had risen fifteenfold. The slave trade continued until the British banned it within the British Empire in 1807.[6] However, money cowries from the Maldives continued to be used to buy red palm oil from West Africa, which was the fuel used during the Industrial Revolution to light homes and factories.

In the mid-nineteenth century, German traders started to use the ring cowrie (*Monetaria annulus*), a species with a similar shell but slightly larger and more pronounced gold ring on its dorsum than its cousin the money cowrie, to purchase slaves in West Africa. This new shell, sourced in East Africa, was cheaper and closer to home than the money cowrie from the Maldives. In a short time the new cowrie became adopted and used together with the money cowrie. During the height of the shell trade, cowries were the most widely used form of currency, but hyperinflation made them inefficient, so they started to be replaced by metal coins. Eventually the value of cowries dropped significantly, and the markets were flooded with them, leading to the collapse of the shell trade by the early twentieth century. By then, the Maldives had exhausted supplying money cowries, after some six hundred years of the shell trade. In all, about 30 billion cowries were collected in the Maldives.[7]

Shell Money in the Americas

Native Americans used several shell currencies for centuries for trading and as a form of gift exchange, but they were commoditized only once the Europeans became involved in trade. In the northeastern United States and Canada, beads were made from clam shells called *quahog* – a word of Narragansett origin, meaning 'clam' – also known as the northern quahog (*Mercenaria mercenaria*, family Veneridae). This clam has a thick shell that is mostly white inside, with a smaller dark purple area. Purple beads (*saki*) were more valuable than white or yellowish beads (*wampi*). The beads were often shaped as long cylinders, about 6 millimetres (¼ in.) long and 3 millimetres (⅛ in.) wide, with a hole in the centre. More rarely, much larger beads were produced but they were more difficult to make and few people had the skills to produce them. As previously discussed, *wampum* beads were strung using fibres or leather as necklaces, bands or as belts. *Wampum* belts were used in diplomatic agreements between Native American tribes, and later, with Europeans.

Among the other shells used to make *wampum* were the large gastropods *Busycotypus canaliculatus* and *Busycon carica* (family Busyconidae), which yielded white beads. Besides being used to make beads, northern quahogs were and continue to be very important as food; this clam is one of the species used to make clam chowder in the New England region. A related species, the ocean quahog (*Arctica islandica*), is also used for clam chowder and clambake, but it was not a source of *wampum*.

Money belt made of beads and Dentalium, *Athabascan people, Tanana, Alaska.*

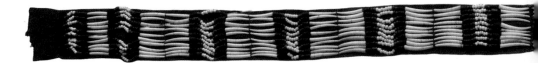

As previously mentioned, on the Pacific coast, *Dentalium* or tusk shells were also used as a form of shell currency.

In western Mexico, spines from the spiny oyster (*Spondylus crassisquama*, family Spondylidae) were valued and fashioned into beads used for trading.[8] The beads were called *Spondulix*.

The Chumash Indians on the Channel Islands, off California, produced the shell money *anchum* from small discs cut from shells of the purple dwarf olive (*Callianax biplicata*, family Olividae) and strung them into strands of beads. The name Chumash originally meant 'bead money makers', and the value of a strand of beads was measured as how many times it wrapped around one's hands. In the Andean region, shells of the Panamanian pearly oyster (*Pinctada mazatlanica*, family Margaritidae) were used as a trading currency.

Other Shell Money and Trivia

Strings or necklaces with shell money have been found in many cultures because they were a convenient way to carry currency. Besides the examples provided above, shell bead necklaces were produced and used as money in West Africa and elsewhere in Africa. Clams of different species were shaped as discs of similar size, drilled and strung as necklaces. Similarly, in Tibet, thick discs and rounded beads were carved from sacred chank shells and strung into necklaces.

In the Solomon Islands, shells of the carnelian olive (*Oliva carneola*, family Olividae) were turned into beads and used as currency in trade. In northern Australia, various tribes made money from different shells; often, currency from one tribe was not accepted by another tribe. In Benguela, West Africa, shells of a relative of the giant African snail (*Lissachatina fulica*, family Achatinidae), *Achatina balteata* (previously *A. monetaria*), would be cut into circles and used as currency. In the Aegean Sea, bracelets, bangles and ornaments made from the shells of *Spondylus* sp. were traded throughout Europe and the Middle East as long as 5,000 years ago.

Clam (or *clamshell*) is an American slang term for money (among over fifty other terms). It is derived from the use of shell currencies. Other slang terms related to shells include *spondulix* (or *spondoolicks*), derived from the *Spondylus* shell currency,

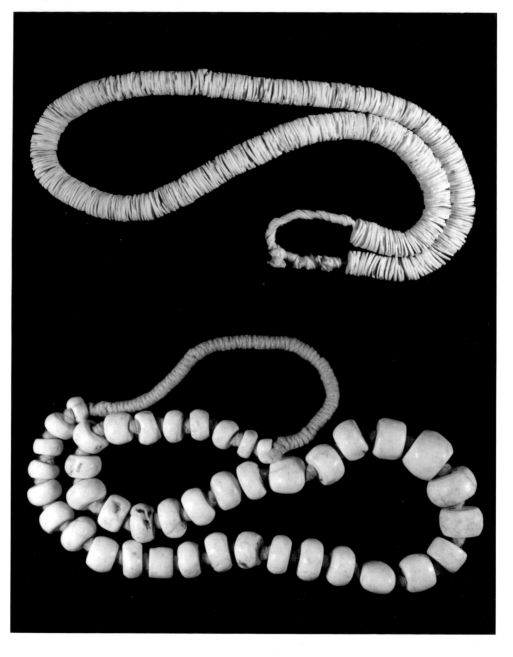

and *wampum*, from the shell beads made from clams in eastern America. Also, the expression *shell out*, meaning 'to disburse or pay out', has its origin in shell money.

In classical Athens (and potentially other cities of ancient Greece), around the fifth century, once a year citizens would carve on oyster shells or pottery shards the name of someone powerful and potentially tyrannical or unwanted. The shells or shards (*ostrakon*) were deposited into a pot and later counted. Upon receiving enough votes, the person would then be banned from the city for a period of ten years. This practice, known as 'ostracism', was used to exile a person through a democratic vote without a trial or specific accusation; no defence was possible. The person retained their property and, once the ten years had passed, they were allowed to come back and join society with their reputation and property intact.[9] This mechanism allowed for a peaceful banning of emerging tyrants.

A unique cheque was drawn on a large clam shell from the west coast of the United States in 1904. The cheque was honoured by the bank, and the shell remained on display there for many years, until it was reportedly stolen and never returned.[10] Because of its uniqueness, the clam cheque was worth more than the amount drawn (U.S.$5).

Shells on Banknotes and Coins

Shells have featured on dozens of banknotes and coins in many countries. In fact, there are some shell collectors who specialize in collecting currencies with shells. Seashells are depicted more frequently than non-marine shells, so they are more common in coastal countries. Some examples of shells on banknotes include Bhutan's one-ngultrum banknote, which features a dragon and a chank shell, and the ten-ngultrum banknote, which also features a chank; Cambodia's one-hundred-riels banknote, featuring a scallop motif; Cayman Island's one-dollar bill, which depicts three gastropod shells (interestingly, the shells appear as dextral on one side and sinistral on the reverse); Cook Islands' three-dollar bill (an unusual denomination), with a top shell on one side and a string of shells on the reverse; French Overseas Territory's five-hundred-franc bill, with a trumpet shell and a spider conch; the Maldives' twenty-rufiyaa bill, with the money cowrie; Papua New Guinea's five-Kina bill, which illustrates a *kina* (pearl oyster) shell; the Philippines' 1,000-peso bill, with a pearl oyster; the Seychelles' ten-, twenty- and 25-rupee banknotes, with several cowrie shells; and Sri Lanka's twenty-rupee banknote,

Necklaces were a convenient way to carry shell currencies. The top is an example from West Africa made from clams, and the bottom necklace, from Tibet, is made from sacred chank beads.

with nautilus and murex shells (curiously represented in portrait orientation, while the reverse is in landscape orientation), among others.

Examples of coins that illustrate shells include Bahama's one-dollar coin, depicting a queen conch (*Aliger gigas*), in great detail; Bhutan's ten-chetrum coin, with a sacred chank (*Turbinella pyrum*); Ghana's one- and twenty-cedis coins, which feature a rat cowrie (*Trona stercoraria*); Jersey's twenty-pound coin, with an abalone shell (*Haliotis* sp.); and Tuvalu's one-cent coin depicting a spider conch (*Harpago chiragra rugosus*), while the fifty-cent coin features an octopus (*Octopus* sp.). In the Pacific Ocean, trumpet shells (*Charonia* spp.) are featured on several coins, including Cook Islands' five- and twenty-dollar coins; the Republic of the Marshall Islands' one-dollar coin; and Vanuatu's two- and five-vatu coins, among others.

Postal stamps are not legal tender in the USA and UK, but once upon a time, they could be used to pay for goods – for example, during the United States Civil War, when there was a shortage of banknotes and metal coins, Congress passed a law that permitted the use of postal stamps as currency.[11] Therefore, postal stamps are included in this chapter on shell money. There are thousands of stamps from all over the world that depict shells. In fact, there are many shell collectors who also collect stamps; some specialize in stamps that depict shells. Perhaps the most comprehensive catalogue of stamps that feature shells is that compiled by Tom Walker, which lists over 5,000 stamps (with details

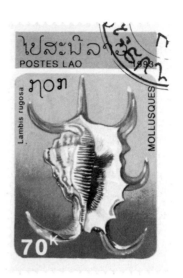

*Opposite top: This banknote for 20 rufiyaa from the Maldives appropriately depicts the country's original currency, the money cowrie (*Monetaria moneta*). The Maldives were the main source of the billions of money cowries collected from the 16th to the 19th centuries. Opposite bottom: One-dollar coin from the Bahamas, depicting a queen conch. Right: Shells are used on over 5,000 stamps around the world. This one is from Laos and shows a spider conch (*Harpago chiragra rugosus*).*

shown for more than 1,500 of them) illustrated with recognizable molluscan shells or parts of shells (for example, mother-of-pearl).[12] However, the catalogue does not include pearls if illustrated as jewellery, and it only includes officially issued stamps.

A small sample of shells on stamps from different countries includes Brazil's 1989 stamp showing a Matthews's morum (*Morum matthewsi*); Burkina Faso's 2000 stamp depicting a giant African snail (*Achatina achatina*); several Cuban stamps of 1973 depicting different species of *Liguus* snails, for example, *Liguus fasciatus fasciatus*; French Polynesia's 1988 stamp illustrating a bivalved gastropod (*Berthelinia* sp.); a 1993 stamp from Laos depicting a spider conch (*Lambis rugosa*, now *Harpago chiragra rugosus*); Malagasy Democratic Republic's 1992 stamp illustrating a paper nautilus (*Argonauta argo*); Nicaragua's 1988 stamp depicting a West Indian fighting conch (*Strombus pugilis*); an undated stamp from St Helena illustrating a purple snail (*Janthina janthina*); a u.s. 2004 stamp showing a triton's trumpet (*Charonia tritonis*); an undated stamp from Vanuatu showing a tapestry turban (*Turbo petholatus*); and Wallis and Futuna's 1962 stamp illustrating a Venus comb (*Murex pecten*), among thousands of other shell stamps. Some shells are economically or culturally important, or simply aesthetically pleasing, and thus appear on multiple stamps, for example, queen conch and other strombs, sacred chank, nautilus, oysters, cowries, cone shells, scallops and so on.

CHAPTER FIVE

HOW THE COWRIE GOT ITS SPOTS

Cowries, marine gastropods in the family Cypraeidae, are diverse and occur in tropical waters worldwide. They have been admired and used by humans for many purposes, including as amulets, jewellery, tools, currency, objects in religious rituals and more. It has been suggested that shell collecting started with cowries,[1] and today they rank among the most popular and best-known groups of shells.[2] Virtually any seashell collection that is not specialized in a family is likely to have at least some cowries, particularly given the lengthy relationship that humans have had with them, but they are showcased here as an example of the diversity and beauty of shells.

The cowrie shell is different from most gastropods. While the typical gastropod has an asymmetrical, coiled or conical shell, the adult cowrie has an inflated or elongated oval shell that looks bilaterally symmetrical on the outside; on the inside, however, the shell is still coiled and asymmetrical. Cowries start out as typical gastropods, and both larval and juvenile cowries have the usual coiled shells, but once the animal reaches sexual maturity, the shell stops growing in length and the body whorl turns and encloses most of the aperture, leaving only a long, narrow opening where thickened lips with 'teeth' develop around the aperture. The shell becomes thicker with age, and calluses may develop on the sides or extremities. In most species, the larval shell is covered by an apical callus and not seen in adult shells.

The mantle is the part of the mollusc that secretes the shell by adding calcium carbonate, a proteinaceous matrix and some pigments to the aperture of the shell.

Some common and rare cowries illustrating the diversity of shapes,
sizes, colours and patterns in the family Cypraeidae.

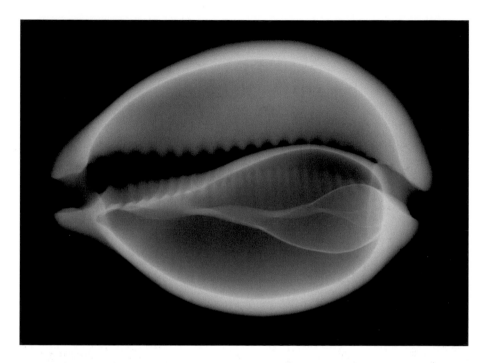

Opposite: Engraving depicting several live cowries and their allies, from J. G. Wood,
The Illustrated Natural History *(1871), vol. III. The four shells in the middle belong to the*
related families Triviidae (false cowries) and Ovulidae (egg cowries). The shell at the top
and the two at the bottom belong to the family Cypraeidae. Above: X-ray of a cowrie shell
showing the adult shell with thickened lips and teeth in the centre and the thin juvenile
coiled shell inside.

Unlike most gastropods, the mantle in cowries has evolved into two large flaps that
completely envelop the shell when undisturbed; when disturbed, the mantle retracts
and exposes the shell. The region where the two lobes of the mantle meet about mid-
dorsum is called the dorsal line, and it usually has a different colouration. The dorsal
line can be curved (as with *Cypraea tigris*), straight, meandering (as with *Leporicypraea
mappa*) or exhibit other profiles; it often has a distinct colour, but sometimes it is
inconspicuous.

Besides secreting the shell by adding shell material to the aperture, the cowrie
mantle also lays dozens to hundreds of thin lamellar layers of translucent or pigmented

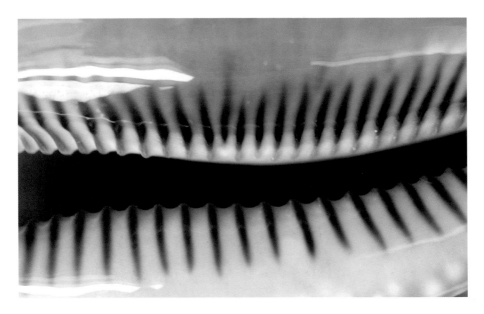

Above: Close-up of the aperture of a cowrie showing labral and columellar teeth.
Opposite: Two cowries, Leporicypraea mappa *(left) and* Cypraea tigris *(right).*
Leporicypraea mappa *is an apt name because its meandering dorsal line resembles*
a map of a coastline.

calcium carbonate. The result can be a stunning three-dimensional colour pattern, as in *Barycypraea fultoni* or *Zoila decipiens*,[3] or simply a diaphanous pattern as in other species. Such three-dimensional patterns are rare among molluscs, because most shells are opaque as a canvas and display only two-dimensional patterns. In many cases, two-dimensional patterns can be successfully modelled by computer algorithms.[4] It is worthwhile to note that three-dimensional patterns do not correspond to the three-dimensional shape of the shell but a complex colour pattern akin to colourful translucent glass sculptures. A few other shells may display similar three-dimensional patterns, including some olives such as *Oliva porphyria* (family Olividae) and possibly a few margin shells (family Marginellidae).

Because the mantle protects the shell against encrusting organisms and there is no periostracum (a proteinaceous layer that looks like a brownish skin that covers some shells), the cowrie shell is glossy; it is also often colourful, and typically adorned with spots, stripes or other patterns. The cowrie mantle tends to have many projections, called papillae, which are involved in respiration, camouflage and sensory input, and may possibly also act to deter some predators. However, since cowries are not toxic, it is unlikely that mantle colouration is used as an aposematic warning to predators.[5] Papillae vary in colour, number, size and structure. They can be simple and finger-like or branched and complex.[6]

Cowries do not have opercula to close their apertures, as many gastropods do. Instead, when disturbed, the entire animal withdraws into the shell. The long, narrow aperture prevents most predators from entering the shell, and the teeth provide reinforcements to prevent peeling of the aperture. The cowrie shell is quite strong and resists being broken or peeled by most predators. However, gastropods and their predators, such as crabs, are always in an arms race to develop a strong shell that is not too heavy to carry around; in response, crabs develop large claws that can break said shell. In other gastropod groups, spines, nodules or other textures may develop to increase strength and protect the shell from predators, but cowries are smooth and lack spines; instead, they may evade some predators by having a slippery shell, making

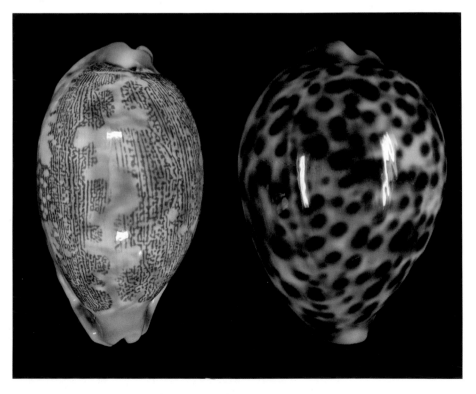

their handling by the predator difficult. However, lobsters, crabs, rays and other fishes sometimes succeed in breaking cowrie shells. The octopus is another cowrie predator, but instead of crushing the shell, it drills a small hole by using its rasping radula and a salivary gland that produces an acidic mucus. Once the hole penetrates the shell, the octopus injects a poison that paralyses and kills the cowrie animal, and then proceeds to eat the flesh. Recent research demonstrates that octopods learn how to drill the boreholes near the attachment site for the columellar muscle; once the muscle detaches from the shell, the animal can be easily removed.[7]

Like other gastropods, cowries have a rasping organ, the radula, that they use to scrape algae or sponges. All cowries have a taenioglossate radula, which has seven teeth per row, but the number of rows, size and shape of teeth vary from species to species.[8] Radulae are used as an important character in gastropod taxonomy.

Radula of Cribrarula gaskoini *under the scanning electron microscope (SEM). This radular ribbon is about 0.152 mm wide.*

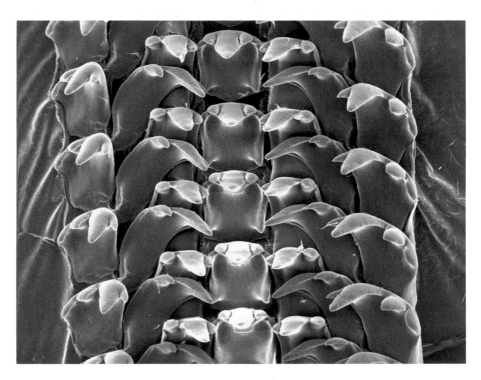

Cowries are also unusual among gastropods because the female shows strong parental care behaviour, guarding her eggs until they hatch. Most gastropods simply lay their eggs and abandon them to fate, but female cowries have been reported to stand guard on or near the eggs, and actively protect them from intruders. Therefore, never disturb a cowrie that is found near an egg mass; this will give them a chance of surviving.

Spotted Cowries

The sieve cowrie (*Cribrarula cribraria*) has a shell with a striking and easily recognized pattern: an orange-brown dorsum with many white spots, and a white base. When the animal is undisturbed, the red mantle covers the entire shell, and it blends in with the red or orange sponges on which the cowries feed. The mantle is semi-translucent, and the white spots on the shell can be seen through it. The mantle has many finger-like papillae located right in the centre of each white spot on the shell. This is not a coincidence; the dorsal spots are formed under each papilla.

When the animal grows and approaches maturity, the shell changes from the typical spiral shell of the juvenile to that of an adult cowrie. Then the mantle secretes a series of thin, orange-brown pigmented layers, except under the papillae. The best explanation is that the microchemical conditions under the papillae do not reach a threshold to trigger the deposition of pigmented layers.[9] The pretty spotted dorsal pattern in the sieve cowrie is the result of a series of pigmented layers with holes that show the white shell underneath; the pattern is indeed a sieve, and as we shall go on to see, Carl Linnaeus could not have found a better name when he described it in 1758 (*cribraria* means 'sieve' in Latin). This is how the cowrie got its spots (at least in this species; there may be a few different ways other spotted cowries develop spots on their shells).

The mantle of cowries is stationary in relation to the shell, meaning that although the animal moves the mantle and retracts it every now and then, when fully extended it returns to the same part of the shell. This allows for the colour pattern to be laid in a precise way most of the time; in those cases, the colour patterns may be sharp dorsal spots, as in most species of *Cribrarula*. However, some species have naturally 'fuzzy' spots, without well-defined margins. In other cases, small disturbances or accidents may result in dorsal colour anomalies; those are usually treasured by shell collectors. Some examples of unusual colour patterns occur in species of *Cribrarula*, such as 'double-exposure', which results in two or more sets of spots slightly offset; rings around dorsal spots; hyperpigmentation or the opposite, lack of pigmentation; and other anomalies.

Above: Two live specimens of Cribrarula cribraria *at Kwajalein, Marshall Islands.
The animal on the left has the mantle mostly withdrawn, and the one on the right has
the mantle fully extended, completely enveloping the shell. The dark orange dorsum with
white spots can be seen through the mantle. Note the finger-like projections of the mantle
(papillae). The two tentacles are seen at the extremity on the right.
Opposite: Flamingo tongue snail (*Cyphoma gibbosum*) and shell.*

As many as 30 to 40 per cent of cowries have spots, dots, eyes and other
designs on their shells that may be related to the number of papillae on their mantles,
but the mechanism of spot formation likely varies from species to species. At least
in the sieve cowries and related species in the genus *Cribrarula*, which was the topic
of this author's PhD dissertation research,[10] there is a one-to-one correspondence
between the number of spots on the shell and the number of papillae on the mantle.
Therefore, shell colour and dorsal pattern in cowries, often dismissed as too variable,
may be more informative than generally regarded, and provide clues about the
animal. The dorsal spots represent more than just pretty designs. Although rarely
used to distinguish cowries, the characteristics of the mantle and papillae, as
well as other anatomical details, may be useful as taxonomic characters to tell
species apart.

All fifteen or more species in the genus *Cribrarula* have somewhat similar spotted shells, but they vary in shape and size; number of dorsal and marginal spots; dorsal colour; number of columellar and labral teeth; features of the dorsal line; and so on. In most species, each feature varies within a narrow range, with some overlap, but a combination of conchological features can separate the species without much trouble. Molecular data have also been used to characterize most species in the family Cypraeidae.[11]

Shell Colour

The colour of shells is one feature that has most captivated people. In some cases, the colour of the animal matches that of the shell, as in the gold-banded latirus, but in other cases, the animal has very different colours than the shell, such as the flamingo tongue (*Cyphoma gibbosum*, an egg cowrie in the family Ovulidae); the animal has striking orange spots lined with black against a yellow mantle, but the shell is typically cream, lacking designs. Why are shells coloured? Nobody knows for sure. There are several theories, but they haven't been tested to satisfaction. One of the mysteries that puzzle malacologists is why some molluscs have colourful shells covered by a

thick periostracum or have coloured parts inside their shells, which cannot be seen while the animal is alive, suggesting that colour may not be used for communication. Another theory proposes that shell colour is associated with camouflage, and that it may give more protection from visual predators. Certain species of *Simnia* (family Ovulidae), which live on whip coral, feed on the polyps of its host, and incorporate the pigments into their shells. Transplant experiments show that the shell of a yellow simnia transplanted to a purple host will continue growing with purple pigments. Some explanations of shell colour advance the idea that shell colours are the result of metabolic processes and that some molluscs deposit pigments in the shell to be rid of certain wastes. This may be a reasonable explanation in some cases, but not all, otherwise molluscs with unpigmented shells would die from metabolic poisoning.

A population of *Cribrarula cribraria* from the Great Barrier Reef region in eastern Australia, *C. cribraria melwardi*, has specimens with varying degrees of non-pigmented (white) shells, but the animals have the same bright red colour as their normally coloured shelled cousins. Some of that population have shells that have no pigmentation (and are sometimes referred to as '100 per cent albinistic'), while others are fully or partially pigmented, with only patches of the normal pigmented layer. Non-pigmented shells of this species are rarely found elsewhere, but they do exist. There are no signs that the animals with white shells are less fit than their normally coloured brethren.

On the other end of the spectrum, some cowrie specimens are more heavily pigmented than normal, as found in a phenomenon known as melanism. A few populations around the world may have a higher incidence of melanistic specimens. Prony Bay in New Caledonia is one of the more famous places to find melanistic cowries; there it may affect almost forty species of cowries, although only a few specimens in each. It is related to the extensive nickel-mining operations in the area. Curiously, some specimens also exhibit deformities and grow shells with elongated extremities or bent and curved shells; these are known as rostrate. The darker and more deformed the shell, the more desirable and expensive it becomes to collectors. One wonders when someone will succeed in developing a technique to raise melanistic and rostrate specimens in captivity.

Cowrie Taxonomy

Carl Linnaeus proposed the binomial system of animal classification in 1758, and the hierarchical system continues to be used today – with some improvements, certainly, but few modifications. Linnaeus was a Swedish naturalist who did not travel much, but his students and specimen collectors travelled the world and brought back many exotic plants and animals for him to study. Among the thousands of animals named by Linnaeus in the twelve editions of his *Systema naturae*, he described the genus *Cypraea*, which included all 39 cowries he named (plus the first synonyms, and one species now in the family Triviidae). His cowries included some of the most widespread, most common and largest species in the family Cypraeidae.

In the following century, as navigation improved and world commerce increased, shells and other natural history specimens became popular as curios in Europe. The desire to have the most exotic specimens ushered in the Golden Age of Discoveries in the early nineteenth century. The British malacologist John E. Gray, a prolific author, named many cowries (among other molluscs), causing the number of species to triple between 1800 and 1900. He also proposed new genera to accommodate the growing diversity of cowries. Other malacologists continued to add more species and genera to the family, and genera were split into increasingly smaller groups, to the point that in 1946 there were 61 genera in the family, including may monospecific genera, such as the genus *Cypraea*, which now included only *Cypraea tigris*, the type species of the family.

Dr E. Alison Kay of the University of Hawai'i studied the anatomy of cowries for her PhD research project and concluded that most species were conservative. As a reaction against the unjustified oversplitting of generic names, she proposed in 1957 to return to the use of the single Linnaean genus *Cypraea* until more information could be obtained. Many researchers and authors of popular books on cowries followed her advice, and amateurs rejoiced, as they only needed to learn a single genus. However, another group of authors continued to employ a multigeneric classification. Today, some sixty years later, we have a great deal more information available. Advances in DNA sequencing in recent decades, allied to increased morphological data, allow for more comprehensive classification, and most authors seem to have adopted the multigeneric approach. Having species arranged in phylogenetic groups has the advantage of keeping most closely related species together, which also provides predictive powers.

There are currently about 250 species and over 160 subspecies of cowries recognized,[12] in addition to many varietal forms that are recognized only by collectors. Although most cowrie species have already been discovered, occasionally new species are still described, often from remote locations but, sometimes, hidden in plain sight. More commonly, however, variations in geographically isolated regions are regularly proposed as new subspecies. Cowries have fewer shell characters than most gastropods:

the larval shell (protoconch) is overgrown by shell material in the adult cowrie, and there are no spines and few sculptural structures that are useful to distinguish cowries, except for the dentition on the thickened lip around the long aperture. Hence, colour patterns, shell size and shape, and apertural teeth are used as the main taxonomic characters. As in other gastropods, radulae are also an important taxonomic character. Over the last half century, anatomical and then molecular characters have been used to provide a more objective and rigorous approach to cowrie taxonomy, but the popular literature is still based heavily on conchological characters, which may bring problems, as some conchological characters may be more plastic than molecules or anatomical characters.

There is a large body of scientific and, especially, popular literature on cowries. Their distribution and range of variation are well documented in many good popular books,[13] articles and, more recently, a magazine dedicated to cowries, *Beautifulcowries*. However, despite the great popular interest in the family, the editor of *Beautifulcowries*, Marco Passamonti, recently made the argument that cowries have received relatively little attention by scientists (compared to other shell families, and in contrast to the popular literature), and called them 'neglected animals'. Hopefully, the beauty of these fascinating animals will also inspire and attract the attention of more scientists, so that we can better understand them and thus help protect them for future generations to enjoy as living creatures, rather than beautiful shells in collections.

Humans and Cowries

Cowries have also been associated with fertility in several cultures, probably because the aperture of the shell is reminiscent of female genitalia. There are several symbolisms related to cowries, including it being used as a charm for fertility by young women in certain tribes in Africa. In other places, cowrie shells were given to women to hold in each hand during childbirth as an amulet for fertility and an easy delivery. This author's mother-in-law experienced it when she had her first child in Palau, Micronesia, in the mid-1960s. Since Palau was colonized by the Japanese in the early 1900s, the custom is likely borrowed from Japan, where it is known as *koyasugai* (easy birth). A similar account is known from West Papua.[14]

The word 'cowrie' (or cowry) in the English language is derived from the Hindi and Sanskrit words *kauri* and *kaparda*, respectively, and related to the Tamil word *kōtu*, meaning 'shell'. The genus *Cypraea* was coined by Linnaeus after Cyprus, the island of Aphrodite, the goddess of love in Greek mythology, an allusion to the shell's aperture resembling female genitalia. The word 'porcelain' comes from the smooth and glossy appearance of the cowrie shell; it has its origin in the old Italian word *porcellana*, meaning 'piglet' because of the resemblance of the shell to a little pig (another theory states that the aperture of the shell resembles the sow's external genitalia). Cowries are known in France as *porcelaine*, the same word for 'china'. In Japanese, the kanji character for shell (pronounced *kai*) is the same as in Chinese, and it is said to be a simplified drawing of the aperture of a cowrie shell. The Japanese word for cowrie is *takaragai*, meaning 'treasure shell'. In Brazil, cowries are known as *cauri* or *búzio* (the latter is also a generic term for seashells) and used in divination games.

Shell Collecting

Cowries are diverse and come in an array of beautiful and colourful patterns. Their shells are aesthetically pleasing, and feel good in your hand, so it is easy to understand why they are one of the most popular shells among collectors. Cowries also range from abundant to very rare, so there are species to fit every shell collector's skill or budget.

Most species are found in the shallow, warm waters of the tropical Indo-Pacific region and share similar distribution with hard corals; many are found from the intertidal zone to shallow depths, so that even snorkellers can potentially find them, but some species occur in waters as deep as 700 metres (2,300 ft). A few other species live in temperate waters off Italy, Japan and southern Australia.

Often, shell collectors specialize in collecting a particular group of shells, or shells from a certain geographic region. Among cowrie collectors there are some that are ultra-specialized and collect every little variation and rare form. It is not surprising that there are collectors that pay premium prices for particularly beautiful (known as 'gem') specimens. Gem-obsessed collectors are common in Europe, China and the United States.

The desire to collect their own shells from around the world drives many people to become avid divers and travellers. Most collectors, however, who cannot travel or hunt their own treasures, resort to trading or purchasing specimens from shell dealers. This demand creates a market which fuels further demand for a large number of specimens as well as the creation of 'new' names. Unfortunately, this has led to a plethora of unnecessary names that confound the taxonomy, a phenomenon that is also seen in a few other shell families, as well as in butterflies, orchids and other collectable organisms. Collectors typically prefer live-collected specimens to beach-worn shells, so shell collecting may be an additional pressure on living marine resources, although not as devastating as pollution and other threats to biodiversity.

Some shells were once regarded as great rarities because only a few specimens were known at the time, including the golden cowrie (*Lyncina aurantium*; the accepted genus today is *Callistocypraea*), It was worth a small fortune in the nineteenth century, but now one can be obtained for less than u.s.$50.

One of the modern rarities, considered one of the world's most desirable shells, is *Sphaerocypraea incomparabilis*. The species name describes it well: a remarkably beautiful and distinctive shell that cannot be compared to any living species. It is, however, very similar to the shells of fossilized molluscs that lived 20 million years ago. This species was originally described as a cowrie, but the World Register of Marine Species currently places this taxon in the family Ovulidae.[15] Six specimens were dredged in deep waters off Somalia by a Russian trawler in 1963 (a seventh was reportedly found but then tossed into the ocean after a dispute over ownership). In recent decades, the area has become dangerous due to pirates, therefore no new specimens have become available, making the six known specimens highly coveted. One of the shells was stolen from the American Museum of Natural History and sold to a private collector in Indonesia for u.s.$20,000. Since the shell is so distinctive, it was easily recognized, recovered and returned to the museum. Perhaps one day, when Somalia is safe again, fishing boats can return to the area and more shells will come to light, but until then, the allure of an impossible-to-get shell will persist.

Sphaerocypraea incomparabilis, *one of the rarest and most desired shells in the world. This is the holotype at the Muséum national d'Histoire naturelle, Paris.*

'A PEARL IS A LIVING JEWEL'
MIKIMOTO KŌKICHI

Pearls are the only gems produced by animals. Because they are unique in that they are naturally glossy, iridescent and do not require to be cut or polished to be appreciated as gems, they were the first objects to be treasured as such, dating back thousands of years. Some of the earliest examples of pearls found in archaeological digs are from Mesopotamia, eastern Arabia and Iran. Later examples are more common and more widespread, and pearls appear in numerous early texts, such as those of Pliny the Elder, who commented in his *Historia naturalis* that pearls were the most popular gem in the Roman Empire in the first century CE.

Pearls are concretions of calcium carbonate deposited in nacreous layers by the mantle of a mollusc to entomb a foreign particle or small organism that enters its mantle cavity. After an initial irritation, if the mollusc cannot expel the intruder, mantle cells start to secrete hundreds of thin concentric layers of calcium carbonate in the form of flat, hexagonal crystals alternated with a thin organic matrix of flexible fibres, called conchiolin, that hold the crystals like bricks in mortar. The final structure is mechanically quite strong to compression, yet pearls can be damaged, and the layers can be peeled, much like an onion.

Right: Head of Venus with a pearl earring, Roman, c. 100 CE, bronze with gold and pearl. Opposite: South Sea cultured pearl necklace.

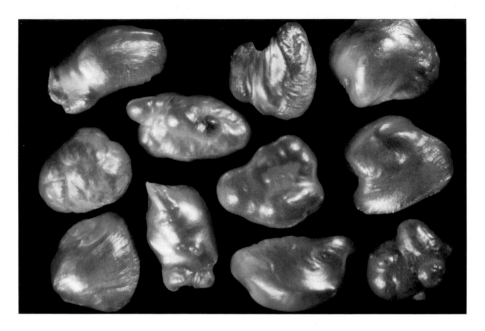

In nacreous pearls, the crystals are aragonitic, as is the inner layer of the shell of the mollusc that produce them, which is often referred to as 'mother-of-pearl'. Pearls formed by molluscs with calcitic shells are composed of calcitic crystals. The layers vary in thickness and clarity; some are clear, while others may be translucent, or contain pigments or impurities. Differences in layer thickness and chemical composition affect the lustre, iridescence, orient (the faint play of colours on the surface) and colour of pearls. These characteristics, together with size and shape, determine the quality and price of pearls. It is important to note, however, that perfect pearls are a human concept; by their nature, pearls have imperfections.

Pearls can occur in almost any mollusc, but they are more commonly found in bivalves and rarely in gastropods, in part because bivalves tend to be filter-feeders, which increases the chances of particles or parasites entering the mantle cavity. Also, gastropods in general are more active than bivalves, and may expel growing pearls more easily. Pearls have been found in many molluscs, including gastropods: queen conch (*Aliger gigas*, which produce attractive pink pearls made of calcite); baler shells (*Melo melo*), abalone (*Haliotis* spp.), horse conch (*Triplofusus giganteus*, which may also produce pink pearls), cowries (for example, *Macrocypraea cervus*); and bivalves, such as oysters (for example, *Crassostrea virginica*; these pearls are not nacreous but porcelaneous), giant clams (*Tridacna maxima* and *T. gigas* produce the largest known, but irregular, pearls) and mussels (for example, *Mytilus edulis*), among others. However, nacreous pearls are only produced by molluscs that have pearly shells. Only two families of bivalves with rich nacreous shells, the Margaritidae and Pteriidae (pearl oysters), small families of marine bivalves with winged shells and the Unionidae, a large family of freshwater mussels, have been and continue to be the primary source of pearls, both natural and cultivated.[1] The cultivated pearls produced by pearl oysters can reach large sizes and be nearly spherical, while freshwater mussels tend to form irregular pearls, called baroque pearls.

Only a few species in two genera of marine pearl oysters are used to produce pearls commercially. The main species are as follows: *Pinctada margaritifera*, the black-lipped pearl oyster, from the Indo-Pacific, has a large shell and may produce large black pearls, up to about 18 millimetres (⁷⁄₁₀ in.) in diameter, known as Tahitian pearls; *Pinctada maxima*, the silver- or golden-lipped pearl oyster, also from the Indo-Pacific, has the largest shell among pearl oysters, and can produce large white to gold pearls up to 18 millimetres in diameter; *Pinctada fucata*, the Akoya pearl oyster, from Japan and southeast Asia, has a smaller shell and produces smaller pearls ranging from white to yellowish, up to 10 millimetres (²⁄₅ in.) in diameter; *Pinctada mazatlanica*, the

Opposite: Natural pearls from freshwater mussels.
Below left: 'Peacock' black Tahitian pearl from Pinctada margaritifera, *15.9 mm.*
Below right: Golden South Sea pearl produced by Pinctada maxima, *14.2 mm.*

La Paz pearl oyster, from the eastern Pacific, can produce pearls up to 10 millimetres in diameter; *Pinctada radiata*, the Ceylon pearl oyster, one of the species exploited extensively for natural pearls, found in the eastern Mediterranean, Red Sea, Persian Gulf and Indian Ocean, produces white to yellowish pearls up to 7 millimetres ($^3/_{10}$ in.) in diameter; and *Pteria penguin*, the black-winged pearl oyster, ranging from the Red Sea and Indian Ocean to the tropical western Pacific, produces Mabé pearls up to 25 millimetres (1 in.) in diameter. A few other species are commercially used, but they produce smaller pearls.

Only a handful of species of freshwater mussels are exploited commercially to produce pearls, out of the 1,000 or so species in the family Unionidae (American rivers have the highest diversity of freshwater mussels in the world). Among the main pearl-producing species are *Hyriopsis cumingii*, the triangle shell pearl mussel, from China and Japan, which produces pearls in different colours; *Cristaria plicata*, the cockscomb pearl mussel, also from Japan and China, is the preferred species used to produce Buddha pearls; *Margaritifera margaritifera*, the European pearl mussel, from Europe, northeastern America and northwestern Asia, is the main species for freshwater

Below left: Pendant in the form of a swan, 16th century, gold and pearls.
Below right: Finger ring, 1905–14, irregular freshwater pearls aligned vertically and framed within a border of gold flowers and leaves, mounted on a thin gold band.
Opposite: Pearl farm in operation in Hạ Long Bay, Vietnam.

pearls in Europe; and *Megalonaias nervosa*, the washboard pearl mussel, native to the Mississippi River region of the United States and Canada, was the main source of mother-of-pearl buttons, and is currently the source of pearl nuclei used widely in perliculture. Besides a few other unionid species that may also on occasion produce attractive pearls, most are not exploited commercially because the mussels are no longer common, have thin shells or do not usually produce good-quality pearls. Currently, cultured freshwater pearls are mostly produced in China, while smaller operations exist in the United States and Japan. China produced about 1,500 tonnes of freshwater pearls in 2005.

    Natural pearls are often misshapen, irregular or not very attractive; perfectly spherical and iridescent natural pearls are very rare, and thus worth a small fortune. Many irregularly shaped pearls were incorporated into elaborate figurines or ornaments containing gold, enamel and precious stones that emphasized their unusual shapes, especially during the Renaissance and into the twentieth century. The process of pearl peeling, used since the Middle Ages, has sometimes been employed to remove damaged or irregular outer layers in the hope of revealing a more perfect, although slightly smaller, pearl within.

Today, about 99 per cent of commercial pearls are cultured, and nearly spherical pearls of high quality can be produced in large scale, although Grade A pearls represent only a small percentage (about 5 per cent for Japanese Akoya pearl oysters, and less than 1 per cent in black-lipped pearl oysters from French Polynesia). Before marketing cultured pearls, perliculturists sort their pearls by size, shape, colour and quality. Each year millions of pearls enter the market. Despite the price of cultured pearls being lower than natural pearls of the past, cultivated pearls still maintain a high value and continue to be desired by many around the world; however, they are no longer the exclusive domain of the wealthy. With the advent – and, later, acceptance – of cultured pearls as real gems, pearls have become accessible to nearly everyone.

The name Mikimoto is synonymous with cultured pearls because the founder of the company, Mikimoto Kōkichi, was one of the early developers of perliculture in Japan in the late nineteenth century. The technique is still used today around the world with only minor variations. However, Mikimoto was not the original inventor of perliculture; mostly, he was a brilliant marketer who fought hard to get cultured pearls accepted as real pearls by the gemmological community and the public. Because of his legacy, his company continues to be very successful, and its pearls are considered among the world's finest (although not the largest). Mikimoto's technique was developed by one of his employees, Otokichi Kuwabara, in 1896, and quickly patented by Mikimoto. Improvements by Tatsuhei Mise and others led to the technique that is currently used, which employs a spherical bead made from the shell of a freshwater mussel, wrapped in a small piece of mantle tissue (called a 'pearl sack') and surgically implanted into the mantle of the bivalve. Often the pearl sack is implanted into the gonad to avoid the pearl growing attached to the shell and becoming a blister pearl. By using a large, perfectly spherical bead, the pearl oyster only adds a relatively thin layer of nacre to produce cultured pearls. The thickness of the nacre added and length of time needed to reach commercial size depend on the species of pearl oyster, the location of the farm and the temperature, food and other conditions. The length of time needed for the bivalve to lay sufficient nacre varies per location and species, and ranges from about one year for Akoya pearls; two to four years for Tahitian and other marine pearls; and up to seven years for freshwater pearls.

Opposite: Removing the pearl manually, Hạ Long Bay, Vietnam.
Below: Freshwater mussel with Buddha pearls.

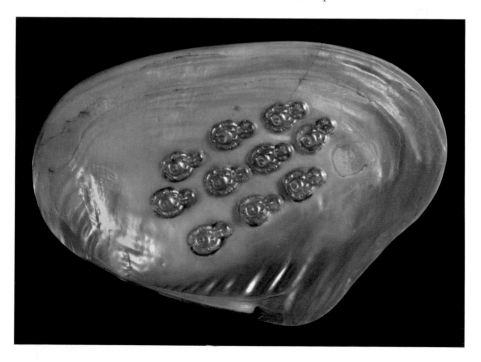

The first known attempts to produce artificial pearls date to the first century CE, as reported by Apollonius of Tyana in the Red Sea. It is likely that the Chinese were the first to successfully produce the cultured pearls (albeit blister pearls) around the fifth century CE by inserting foreign objects into the mantle of freshwater mussels. The technique was eventually refined to the point that blister pearls of almost any shape could be produced, provided a seed of the right shape was implanted in the bivalve. The result was a pearlized object; the most famous ones were Buddha-shaped blister pearls, used for religious purposes.

One of the pioneers of perliculture in Europe was the Swedish naturalist Carl Linnaeus, 'the father of taxonomy', who is best known for creating the system of binomial nomenclature for plants (1753) and animals (1758) (for example, *Pinctada margaritifera*, the black-lipped pearl oyster). He is also credited with producing the first spherical cultured pearl in any mollusc ever in the 1750s. Linnaeus experimented with Swedish freshwater mussels, *Unio pictorum*, and improved upon Chinese methods that produced blister pearls. He drilled a small hole in the mussel shell and inserted a minuscule piece of limestone at the tip of a wire to keep it away from the shell and prevent the formation of a blister pearl. The mussel was then returned to the river, and after six years Linnaeus obtained the first free pearls. The technique was unknown for over a century, until a researcher studying his scientific collections housed in the Linnaean Society of London discovered his pearls and manuscript detailing the method, publishing the latter 144 years after Linnaeus' achievement. Contrary to popular belief, the Swedish crown awarded Linnaeus nobility and knighthood as recognition for his success in perliculture, not for his contribution to botany or nomenclature.[2] Linnaeus sold the rights to his technique to a merchant who was granted a monopoly permit by the king, but never used it to produce pearls commercially.

Besides commercial pearls from marine pearl oysters and freshwater mussels, there is a nascent industry for pink pearls made by the queen conch (*Aliger gigas*, family Strombidae), the Caribbean gastropod with a large shell and a bright pink aperture that is not nacreous. Researchers in Florida have been developing techniques to produce beautiful pink pearls. The pearls are prized but the market is still relatively small compared to the more conventionally coloured pearls. The current obstacle to expanding the market, besides the high price for the conch pearls, is the fact that the attractive pink colour fades over time.

In addition to natural and cultured pearls, imitation pearls were known from the Roman Empire and throughout the Middle Ages, as well as in Renaissance Europe, and continue to be made in modern times. According to several recipes from

*Tiffany & Co. necklace with a pink pearl pendant from the queen conch (*Aliger gigas),
1900–1910, conch pearl, platinum, diamonds. The pearl weighs 3.64 grams.

the fifteenth and sixteenth centuries, including one published by Leonardo da Vinci,
small pearls were ground or dissolved in lemon juice, and the resulting paste dried,
mixed with egg white and formed into larger pearls that were polished. Other early
formulations used ground fish scales and resin, replaced in the modern era by mother-
of-pearl, glass or plastic. A recent technique produces 'shell pearls' from ground pearl
oysters, a binding agent and a dye. The resulting pearls are perfectly round, have

Above: Necklace with pink seashell pearls, a much more affordable alternative to cultivated pearls. Opposite: Ama *with mask and in traditional white clothing, showing an abalone during a demonstration in Toba Bay, Mikimoto Pearl Island.*

the same feel, weight and lustre as natural pearls, but cost a fraction of the price of cultivated pearls. They can, however, be distinguished from natural or cultured pearls when viewed under magnification, on the basis of surface features.

Although natural pearls occur in locations around the world, harvesting pearls by diving dates to approximately 2000 BCE and was generally confined to the Persian Gulf, Red Sea and regions of the western Indian Ocean. Columbus's encounter with pearls worn by the Indigenous peoples when he first arrived in Venezuela led to the establishment of large pearl fisheries throughout the Caribbean coast of South America during the early sixteenth century, which became the primary source of pearls for Europe during the Renaissance.

In Japan, the *Ama* (sea woman) divers were women who dove primarily for seafood but also for other sea life and pearls. The tradition dates to the tenth century. They wore only a loincloth, even in cold waters, to avoid getting clothing caught underwater. In Korea, a similar group of pearl divers were called *Haenyeo*. Involvement of the *Ama* in diving for pearls increased greatly following the establishment of the cultured pearl fishery at the end of the nineteenth century.

Despite the staggering quantities of pearls, both natural and cultivated, that have entered the world's markets over the past two millennia, several individual pearls have achieved notoriety. Perhaps the most famous pearl in the world is La Peregrina (the wandering pearl), a white, pear-shaped pearl collected in the sixteenth century in the Pearl Islands of Panama. At the time it was collected, it was the largest pearl known, weighing 11.2 grams (⅖ oz). It was then sold to King Philip II of Spain, thus becoming a crown jewel, and is evident in the portraits of several queens of the country. Joseph Bonaparte, brother of Napoleon, took La Peregrina to France when forced to leave Spain. His nephew Napoleon III took it to England and sold it to the Marquess of Abercorn. The family then sold the pearl at auction in 1969, where it was purchased by Richard Burton for his wife, Elizabeth Taylor. Following her death, it was again sold at auction in 2011 for more than U.S.$10 million and is now in private hands.

Other notable pearls include La Pelegrina, an exceptionally large, natural round pearl collected off the coast of India in the early 1800s. It weighed 110 grams (4 oz), nearly ten times the weight of La Peregrina. The Pearl of Allah, also known as the Pearl of Lao Tze, has long been known as the largest pearl in the world. It is an

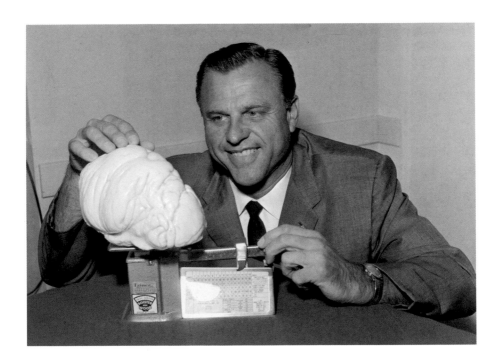

Above: The Pearl of Allah, or Pearl of Lao Tzu, the world's largest and most expensive natural pearl, c. 1964. It measures 24 centimetres (9½ in.) long and weighs 6.4 kilograms (14 lb). Opposite: La Peregrina, a natural pearl, diamond, ruby and cultured pearl necklace, designed by Elizabeth Taylor with Al Durante of Cartier, 1972.

irregularly shaped, non-nacreous pearl produced by the giant clam *Tridacna gigas*, and is said to resemble the turbaned head of Allah. It measures 24 centimetres (9½ in.) in maximum diameter, weighs 6.4 kilograms (14 lb) and was collected in 1934 in the Philippines. Stories of its origin include the drowning of a diver when his hand was caught by the giant clam.[3]

    A far larger pearl taken from a giant clam was recently revealed on the Internet.[4] It is reported to measure 67 by 37 centimetres (26.4 × 14.5 in.) and weigh 34 kilograms (75 lb). It was found by a fisherman from Palawan who kept it under his bed for ten years before turning it in to his village. This pearl has yet to be examined or confirmed by gemmologists, however.

CHAPTER SEVEN

SHELLS IN THE ARTS

What is art? There are many answers to this question, but basically it is the expression of human creativity in its many dimensions, such as painting, music, dance and literature. The delicate spirals, beautiful colour patterns and sculptures of shells have inspired artists throughout history and are incorporated into diverse forms of artistic expression around the world. Some of the early artistic expressions that survived to date include jewellery and shell art used to adorn the living and the dead, shell necklaces found in burial sites, caves with early human artefacts, shell carvings with artistic designs and objects made into the semblance of shells (for example, a cockle shell made of gold from Ur), used for eyeshadow, lamps in the shape of shells and so on. The doodle done on a freshwater mussel shell over half a million years ago by one of our ancestors may not be interpreted universally as art, but it is one of the first expressions of pre-human craft that we know to date. Other examples include ancient inlaid mosaics like those in Pompeii, Italy, depicting shellfish and octopuses, ancient Greek pottery with shellfish designs or even fashioned like bivalves, and more.

Shells in Painting and Print

In the beautiful and intricately detailed nineteenth-century painting of a collection of shells by the French painter Alexandre-Isidore Leroy de Barde, each shell is realistically and accurately represented in shape and colour to the point that most can be identified to species level. Most of the seventy objects in this painting are seashells, in addition to two land snails, one freshwater snail, four sea urchins, two gorgonians, two sponges and one brachiopod shell; one gastropod is shown with a hermit crab in its aperture. This image likely represents the curio cabinet of a wealthy collector who had amassed an impressive collection of worldwide specimens. Cabinets of curiosities were very popular during the Age of Discovery (*c.* 1400–1770) in Europe. As Europeans set sail

Adriaen Coorte, Five Shells Mounted on a Slab of Stone, *1696,*
oil on paper mounted on wood.

to discover and conquer new lands, sailors brought home souvenirs from their distant travels, which sparked the interest of the wealthy and educated. This brought greater interest in collecting more natural history specimens, which then drove adventurers and merchants to find and bring exotic specimens back to Europe. This influx of new specimens led to a golden era of biological discoveries.

The Swedish naturalist Carl Linnaeus, the Father of Taxonomy, did not illustrate his books describing new species but cited illustrations in a considerable number of previously published works.[1] In subsequent years, it became a requirement that new species descriptions include illustrations.

Early scientific works depicting shells are not notable for their artistic quality or the accuracy of the illustrations. In most works prior to the 1700s, shells were drawn as they were seen and then the figures were engraved onto copper plates or other materials, but the resulting prints were sometimes reversed. This is usually not a problem for most natural history specimens, because they are bilaterally symmetrical, but it is a problem for gastropods. Most gastropod shells are dextral: if you point the apex upwards, the aperture opens to the right; therefore, they are known as right-handed shells because you could insert your right hand into the shell and hold its columella. Relatively few gastropods have sinistral (or 'left-handed') shells that coil in the opposite direction. It took time for some artists and publishers to realize the error in reversed prints. Even the great Dutch master Rembrandt produced an illustration of *Conus marmoreus* as a sinistral shell. He had learned how to sign his name in reverse on the engraving plates, so that this signature was printed correctly, but he failed to recognize the reversed orientation of the shell. He later produced another print with the correct orientation. In many instances, artists and printers did not care that illustrations of shells were published as mirror images, or did not think it was important.

Below: Rembrandt's illustration of Conus marmoreus, *1650, engraved correctly but reversed in the printing. Opposite: Illustration of* Murex brandaris, *from Martin Lister,* Historiæ conchyliorum *(1685–92, this edn 1770).*

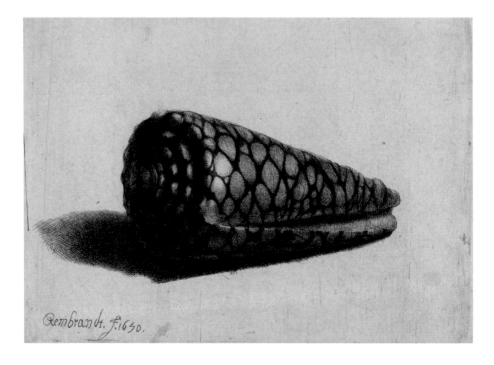

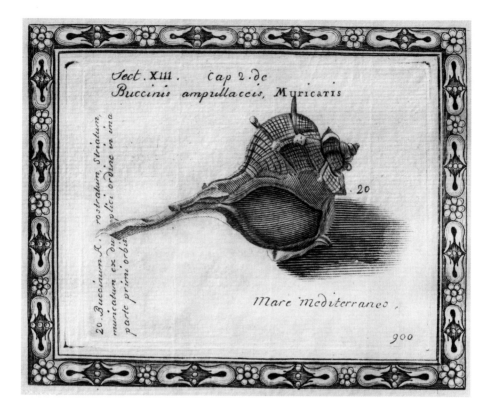

The problem is known as the Gould-Abbott hypothesis, introduced by Stephen Jay Gould and tested by Warren Allmon by looking at many old illustrations of snails and crabs (some of which are also asymmetrical and have one claw larger than the other).[2] Allmon concluded that there is support that the Gould-Abbott hypothesis is correct: artists and printers prior to 1700 may have noticed the reversed images but did not care to produce correctly orientated images of shells.

Over time, artists – or the scientists who hired them – adopted the correct orientation in shell images, and it became the norm following Linnaeus' taxonomy, when naturalists realized that the details of the shell were important in classifying shells. Since then, most shell books have been published with images in the correct orientation, except for a few works in which the printers failed to print images correctly. However, even in recent times, there are occasional books or articles in the popular media that have been published with reversed shell images, even on covers!

This is likely the result of prohibitive costs to replace new covers for an entire print run, or the author not paying close attention to the cover or other plates. The poor author then must be resigned to the shame and ridicule of all who recognize a sinistral shell as an error.

One notable exception among a few early authors who recognized the importance of accurately orientated images was the English physician and naturalist Martin Lister, who introduced the science of conchology to England. His landmark book *Historiæ conchyliorum* (1685–92) illustrated more than 2,000 species of shells, all in the correct orientation. It was the first major shell iconography to be published in the world.[3]

Following the Age of Discovery, not only did scientific discoveries blossom, but the arts flourished in many forms as well. There was a steady progression in the artistic quality and details in shell illustrations to the point of realism. Several naturalists with an artistic bent became accomplished iconographers and produced extensive compilations of drawings and etchings of shells, as well as of other animals and plants. Some superb naturalists and illustrators and their more famous publications are (in alphabetical order) Jean-Charles Chenu (1808–1879, France), *L'Encyclopédie d'histoire naturelle*; Alcide d'Orbigny (1802–1857, France), *Dictionnaire universel d'histoire naturelle*; Niccolò Gualtieri (1688–1744, Italy), *Index testarum conchyliorum*; Louis Charles Kiener (1799–1881, France), *Spécies général et iconographie des coquilles vivantes*; Heinrich Carl Küster (1807–1876, Germany), *Conchylien Cabinet*; Jean-Baptiste Lamarck (1744–1829, France), *Tableau encyclopédique et méthodique*; Martin Lister (1639–1712, United Kingdom), *Historiæ conchyliorum*; Henri-Joseph Redouté (1766–1852, Belgium), *Tableau encyclopédique et méthodique*; Lovell Augustus Reeve (1814–1865, United Kingdom), *Conchologia iconica*; Georg Eberhard Rumphius (1627–1702, Germany), *D'Amboinsche Rariteitkamer*; George Brettingham Sowerby I (1788–1854), G. B. Sowerby II (1812–1884) and G. B. Sowerby III (1843–1921, all from the United Kingdom), *Thesaurus conchyliorum*; Rudolf Sturany (1867–1935, Austria), *Gastropoden des Rothen Meeres*; and William John Swainson (1789–1855, United Kingdom), *Exotic Conchology*.

In paintings, the accuracy in details and orientation of shells was particularly good much earlier than in print. One of the most iconic paintings of shells is *The Birth of Venus* (*Nascita di Venere*), by the Italian artist Sandro Botticelli, completed more than

Alexandre-Isidore Leroy de Barde, Selection of Shells, c. 1803/1810, watercolour and gouache on paper glued on canvas.

550 years ago (*c.* 1484–6). In the painting, Venus rises from a giant scallop shell in the ocean as wind blows over her naked body. This painting has inspired many artists in different areas of the arts.

Carvings and Sculptures

Shell cameos are small carvings made using certain shells that reveal different coloured layers. The artist intimately knows how the coloured layers are formed and carefully carves through one or more of them until the desired colour is exposed. The result is a raised image that contrasts against the background colour. The most sought-after shell cameos come from Torre del Greco in Italy, and the two species of choice are both helmet shells, although a few other species are less frequently used: the bull-mouth helmet (*Cypraecassis rufa*), with an orange-brown inner colour and a whitish outer colour, and the queen helmet (*Cassis madagascariensis*), which is larger, with an alabaster outer colour and chestnut-brown interior. The most common designs depict silhouettes of ladies, monarchs, scenes from Roman and Greek mythology or floral designs. Shell cameo brooches or cabochon are most popular, often with gold accents, but larger pieces including the whole shell, partly carved, are also found. The oldest shell cameos are about 1,000 years old, but they became more popular during the Renaissance. Stone cameos are considered more valuable than shell cameos, but good pieces of the latter

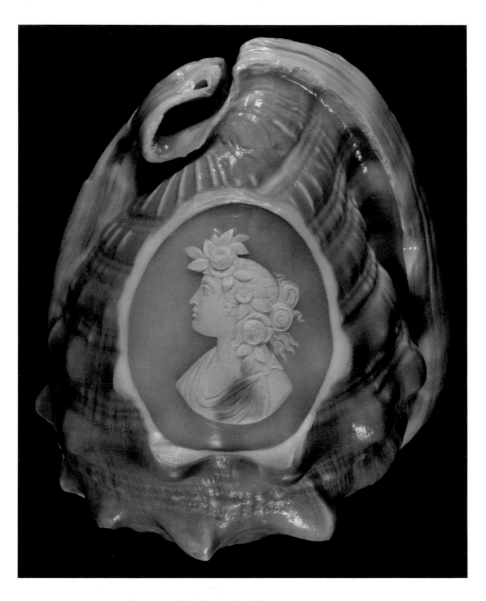

Opposite: Sandro Botticelli, The Birth of Venus, *c. 1485, tempera on canvas.*
*Above: Carved bullmouth helmet (*Cypraecassis rufa*) shell cameo with the silhouette*
of a lady.

can be quite expensive and are treasured as heirloom jewellery. The Italian shell cameo industry is valued at about u.s.$200 million a year and employs thousands of artists.[4]

Among the most iconic sculptures with shells are a pair of white marble sculptures by the French artist Jean-Baptiste Carpeaux (1827–1875). One depicts a full-sized eleven-year-old boy, crouched naked and listening to a conch shell next to his ear: *Neapolitan Fisherboy* (*Pêcheur napolitain à la coquille*, 1857–62); and the other takes the form of a young girl, also naked, with a conch shell on top of her head: *Girl with a Shell* (*Jeune fille à la coquille*, 1863–7). The National Gallery of Art in Washington, DC, has one of the three pairs made by the artist. He also made smaller versions of these sculptures in bronze that were sold to the public.

Netsuke are miniature three-dimensional carvings from Japan, ranging in size from about 3 centimetres (1⅕ in.) to 8 centimetres (3 in.) long, that are attached to the end of an *obi* (belt) worn with a kimono or similar Japanese attire. They come in many shapes and materials besides ivory and wood. Some of them are made in the shape of seashells, snails, octopuses, clams and other shells, besides figures from Japanese folklore. The carvings can be simple or exquisitely detailed, realistic or whimsical. Each is a work of art on its own, which explains why *netsuke* have become collectible items. More affordable reproductions of special ones may be found in museum and other shops in Japan. This art form flourished during the Edo period in Japan (1603–1868) but continues today. More recently, as ivory from living animals has been banned, artists have resorted to ivory or teeth from mammoths and other extinct animals, which are still legally available; different types of wood have always been an alternative material, however. The tagua nut from Central and South America (genus *Phytelephas*, meaning 'vegetable elephant'), also known as the ivory palm or vegetable ivory, produces a nut that closely resembles elephant ivory. It is very hard when dry, but it can be easily carved when

Right: Netsuke *of shell,
19th century, ivory.
Opposite: Jean-Baptiste
Carpeaux,* Neapolitan
Fisherboy, *c. 1857–62, marble.*

wet. It is promoted as a sustainable substitute to ivory, and its size, 4–8 centimetres (1½–3 in.) diameter, is ideal for *netsuke* carving. Tagua nut has been used for over a century to make imitation ivory buttons.

Films, Animated Movies, Printed Media and Music

Shells have featured in many films and animated movies, sometimes appearing just as empty shells, other times as characters, with the living mollusc inside. The following are just but a few examples. In the movie *Dr No* (1962), there is the famous scene that made Ursula Andress famous, where the bikini-clad skin diver Honey Ryder is seen coming ashore with two queen conchs that she has just collected in her hands. Honey is singing, while holding the shells; and her singing wakes up James Bond, played by a young Sean Connery in his first James Bond film, who is sleeping in a hammock nearby. In *Demolition Man* (1993), police officer John Spartan (played by Sylvester Stallone) is sent to the future; after he goes to the bathroom, he comments to his lieutenant, Lenina Huxley (Sandra Bullock), that they were out of toilet paper; she then laughs and comments to the others that he did not know how to use the three seashells in the bathroom. There are several forum threads on the Internet regarding the meaning of the three seashells in this movie: some think the shells activate a bidet; one poster states that they interviewed the screenwriter, who reportedly said that he added shells to the movie as a joke and that there is no logical explanation for them. The award-winning movie *Jilel: The Calling of the Shell* (2014) discusses the dangers of climate change in the low islands of Majuro Atoll, Marshall Islands. In the film, an heirloom trumpet shell with special powers is passed from a grandmother to her granddaughter to help protect their islands against many impending environmental problems faced by island nations, such as rising sea levels.

In the last few decades there have been many animated films with excellent animation and complex storytelling that have included shells, such as *The Little Mermaid* (1989), *Finding Nemo* (2003), *Turbo* (2013), *Finding Dory* (2016) and *Moana* (2016), to name just a few. *Finding Nemo* presents marine animals in a playful way, yet provides mostly accurate facts about them, and the film helped boost interest in marine biology the same way that *Jurassic Park* (1993) sparked an interest in dinosaurs among children a decade before. The producers of *Moana* hired many consultants to avoid making any cultural mistakes; unfortunately, it is clear that no shell experts were consulted, because there are several errors. In the movie the unmistakable pink queen conch endemic to the Caribbean is found in Polynesia, and there are several instances of sinistral shells.

Honey Ryder (Ursula Andress) coming ashore with two queen conchs in the film Dr No *(1962, dir. Terence Young).*

The animated TV series *SpongeBob SquarePants* (1999–), created by American marine biologist Stephen Hillenburg, tells of the quirky underwater adventures of SpongeBob SquarePants, a marine sponge, and a wide cast of friends, including Gary the Snail, SpongeBob's pet. The commercial success of the show (over U.S.$13 billion by 2017) has brought a greater awareness about marine animals, despite its wackiness. Another animated TV series for children is *Shelldon*, a show originally from Thailand, about the quests of Shelldon, a Yoka star turban shell (*Guildfordia yoka*, one of the most distinctive shells, which has a flattened round shell with a few very long radiating spines) and his friends at the Charming Clam Inn; the series includes molluscs, crabs and other marine life.

Comic strips often illustrate shells in newspapers and printed media. *Sherman's Lagoon* (1991–) is a daily comic strip by the American cartoonist Jim Toomey that revolves around Sherman the Shark and his friends in a fictional Kapupu Lagoon in the South Pacific. Because of Toomey's interest in the ocean and conservation, current environmental issues are discussed from a comedic perspective that attracts adults and kids alike. Seashells are often featured. The recent discovery by R. Bieler, T. A. Rawlings and T. M. Collins in 2017 of a new species of wormsnail (*Thylacodes vandyensis*) that uses a mucus network to capture its food was dubbed 'the Spiderman snail' and captivated the popular media. The snail was highlighted in the *Sherman's Lagoon* cartoon strip for several days in January 2018.

Shells are sometimes mentioned in songs, often as treasures found on a beach, as in the Québécois ballad by Ginette Reno and Jacques Boulanger, 'Le Sable et la Mer' (1978; re-recorded by Québécois artists Ingrid St-Pierre and Les Denis Drolet in 2012). The progressive rock band Supertramp released their hit song 'Hide in Your Shell' from the album *Crime of the Century* in 1974, but their best rendition of the song is arguably in their live album *Paris* (1980). It is considered by many as deeply profound, with inspirational lyrics, and the beautiful voice of songwriter and singer Roger Hodgson makes this a timeless classic. The Australian fusion band with both Aboriginal and *balanda* (non-Aboriginal members) Yothu Yindi (Yolngu word for 'child and mother') sings the inspiring song 'Dots on the Shell' (1993) in both Aboriginal language and English. British folk-rock singer Al Stewart has an album named *A Beach Full of Shells* (2005), and in the track 'Somewhere in England, 1915' uses the word 'shells' as a double entendre for seashells and artillery shells, referring to the First World War (the album cover shows both images).

Shells have also been used to make music for millennia. One of the oldest examples is from the Magdalenian culture of the Upper Palaeolithic (about 12,000 BP), found at Marsoulas Cave, France. The most common shell musical instruments are wind instruments, known as shell trumpets, shell horns or conch. As discussed earlier, shell trumpets have been used independently by multiple cultures around the world and for various different purposes, such as religious ceremonies, popular festivities, military exercises and war, and for other occasions. The instruments are produced from large shells such as trumpet shells (*Charonia* spp.), helmet (*Cassis* spp.), conchs (for example, *Aliger gigas*), sacred chank (*Turbinella pyrum*) and a few other large gastropod species (for example, *Bursa bubo* and *Syrinx aruanus*).

Typically, the apex of the gastropod is cut off and the player simply blows into the shell (in end-blown shell trumpets), but sometimes a metal or wooden mouthpart is added. Less commonly, the blowhole is bored on the side of the spire (in side-blown shell trumpets). Many shell trumpets consist simply of the natural, unornamented shell, but in some places – such as Tibet, where it is often seen – the shell may be carved with designs and adorned with semi-precious stones and metals. Most shell trumpets can produce a loud sound but play a single note. Rarely, one or a few fingerholes are bored on the side of shell to provide a change in pitch, or alternatively the player can insert their hand into the aperture to modulate the pitch. A few musicians have incorporated shell trumpets into contemporary music and jazz. Other shell musical instruments include screw flutes, made from gastropod shells with a high spire and multiple finger holes; percussion instruments made from shells; and wind instruments made from ceramic or other materials to resemble a shell, such as an ocarina fashioned as a gastropod shell.

Chank shell trumpet, Tibet, 18th–19th century, made with copper, gilt copper alloy and semi-precious stones.

Jewellery

Shell jewellery has had a long history. The British Museum in London, the Houston Museum of Fine Arts and any major museum with archaeological collections or exhibits demonstrate that the appeal of shells has appeared independently in many cultures across the globe, usually early in each culture. Often, those objects were related to decoration or jewellery, generally among the wealthy or ruling elites, with simpler jewellery among the common folk. Papua New Guinea and other Pacific islands are rich in shell jewellery simply because the area lies within the Coral Triangle, the epicentre of marine biodiversity in the world. The Coral Triangle is a roughly triangular region between the Philippines (north), the island of Borneo and Timor-Leste (west) to Papua New Guinea and the Solomon Islands (east). Even in other regions, shell jewellery can be quite common. In the Indian Ocean and across Africa, cowrie shells are used extensively in jewellery and in textiles. Shells from colder areas typically are not as colourful as those from the

Below: Banjara embroidery with cowrie shells, India. Opposite: Ni'ihau shell leis as necklaces.

tropics. More colourful or nacreous shells may be more treasured, and are, in some cases, reserved for wealthy individuals or the island chief, as we have seen in the case of the golden cowrie in some Pacific islands. Other shells are also used as a symbol of status, for example, *kina* (pearl oysters) shells in Papua New Guinea. The island of Flores, in the Lesser Sunda Islands of Indonesia, has a rich culture with shell necklaces and intricate gold jewellery, with some pieces even making it into museums.

Modern shell jewellery is varied and abundant. Examples may range from simple earrings, bracelets or pendants with a single cowrie or other shells, to fancy and expensive gold jewellery with shells, often the nacre from abalone or mother-of-pearl, or pearls. With the advent of the Internet, small companies and individual artists can today easily sell their creations to the whole world. Fashions that used to be regional now can be obtained by anyone, from anywhere, providing they have Internet access.

Hawaii has a tradition of using seashells (and, less commonly, the colourful endemic tree snails that once were common but are now endangered) to make shell leis, or shell garlands. To many tourists, Hawaiian shell necklaces are the mass-produced, cheap versions sold in Waikiki, or given to visitors at some stores. Most are probably unaware of the traditional and expensive *lei pūpū ʻo Niʻihau*, Niʻihau shell leis, which are considered as fine pieces of jewellery that can sell for U.S.$30,000 per single piece, and once were worn by Hawaiian royalty. They are an important part of the Hawaiian cultural heritage. It is believed that the tradition of making shell leis instead of the more common flower leis evolved on the island of Niʻihau because of the scarcity of flowers,

Left: Queen Emma of Hawaii wearing strands of Niʻihau *shells, c. 1880, photograph by Menzies Dickson. Opposite: Double Sailor's Valentine, Barbados, 19th century, wood, shells.*

due to the arid climate, and abundance of attractive miniature seashells (*pūpū*). Niʻihau is known as the Forbidden Island because it became a private island in 1864 and since then it has been off-limits to all but the relatives and guests of its owners, the Robinson family. It is the smallest of the inhabited Hawaiian islands, at approximately 180 square kilometres (70 sq. mi.) with a population of only 170 native Hawaiians (2010 census). Contrary to popular belief, Niʻihau *pūpū* shells are not unique to the island; they are also found on neighbouring islands and elsewhere, but they are more common on Niʻihau. What makes these shell leis special is the superb craftsmanship that local artisans have developed. To protect these artisans, whose sole income is shell lei making, a law was passed in 2004 reserving the term 'Niʻihau shell lei' only for leis that are 100 per cent made of shells from the island. To make a large lei, artisans may take a year or two to collect shells of the perfect size and colour, and to make the lei. There are several styles and colours of Niʻihau shell leis: some have a single colour per strand, while others combine different species of shells and colours, arranged in patterns. The three main types of shells used for the leis are *kahelelani* (*Collonista verruca*, a turban shell), *momi* (*Euplica varians*, a dove shell) and *laiki* (*Mitrella margarita*, another dove shell), besides larger shells used for clasps, usually a pair of cowries, *pōleholeho ʻāpuʻupuʻu* (*Nucleolaria granulata*, an endemic Hawaiian cowrie with a bumpy shell).[5] The Bishop Museum in Honolulu, Hawaii, has on display several of the Niʻihau shell leis owned by Hawaiian royalty. Queen Emma, one of the most beloved characters in Hawaiian history, was the monarch who treasured Hawaiian crafts the most and wore Niʻihau shell leis on special occasions.

When Queen Kapi'olani attended Queen Victoria's Golden Jubilee, a celebration in 1887 to mark fifty years of the English queen's reign, she wore a multi-strand Ni'ihau shell lei. Today, these leis are worn at special festivities such as a wedding or hula performance, preferably while wearing Hawaiian attire.

Shellcraft and Decoration

Crafts made from shells are diverse and can be found in many coastal areas worldwide. They range from simple figurines made from different species of shells glued together, often sold as souvenirs at seaside tourist destinations, to complex artistic creations. An interesting category of shellcraft is the 'Sailor's Valentine', which consists of intricate designs made from hundreds to thousands of small shells carefully arranged by colour or shape in two octagonal wooden boxes and connected by a hinge, forming symmetric patterns or designs, among them flowers and hearts, often with a message or the name of a loved one. This type of craft became popular in the nineteenth century, when sailors brought these crafts to their loved ones upon returning from long trips. Perhaps sailors originally made these during the idle times on board, but it is unlikely that the seamen had a supply of miniature shells of different colours and shapes at hand to produce the designs, and novelty shops in Barbados and elsewhere sold ready-made crafts to less artistic sailors. Today, antique Sailor's Valentines are treasured, and some are quite valuable. In some American shell shows there are competitions for shellcraft, with a special category for Sailor's Valentines. Some of the contestants take the challenge

Sewing box, Austria, c. 1765, gold and mother-of-pearl.

very seriously and create impressive works of art. Purists use only natural-coloured shells (no dyes!) and mother-of-pearl but no other materials such as wood, glass or seeds.

Other types of craft objects made with shells include picture frames, boxes, crosses, wreaths, wine-bottle toppers, wind chimes and a wide variety of articles decorated with shells or inlaid with mother-of-pearl. Museums, private collections and the homes of both shell enthusiasts and people who live in coastal areas have vast amounts of contemporary and ancient shell-themed decorations. In the United States, decorators are particularly keen on using shells in bathrooms, so there are all sorts of towels, shower curtains, soap dishes, framed prints, door mats, lamp shades and other shell-themed pieces alongside actual shells. It is also quite common to spot shells in home and office decorations in American films and TV programmes – because this author is interested in shells, those details always catch his attention.

The use of mother-of-pearl as inlays in wooden boxes and furniture has a long history in many cultures as diverse as those of Egypt, Turkey, Mexico, China, Japan, Korea, Italy, France, New Zealand and more. The colour varies and depends on the species of the shell or place of origin: pearl oysters come in white, silver, grey, gold and black; freshwater mussels range in colour from the traditional white or silver to pink and brown, while abalone has very iridescent and colourful nacre with varying hues of green, blue, red and pink. The amount of mother-of-pearl varies from marquetry of thin plaques of mother-in-pearl into a wood veneer, to having nearly all the surface of the furniture covered in mother-of-pearl. Often designs use contrasting colours of nacre, bone, ivory, wood or stone to form exquisite geometric patterns, such as those from Islamic art, or flowery designs such as those from France. In Japan, furniture with mother-of-pearl inlay is known from as early as the eighth century, while in France some of the earliest examples are from the sixteenth century.

Inspired by Shells

The artistic reach of shells is not limited to the traditional arts, as mentioned above, but has also touched the culinary arts. Not including the mollusc animal that secretes the shell, which is consumed around the world, but the actual shells have inspired many edible treats. One example is a traditional Mexican pastry, a *pan dulce* (sweet bread with a sugary crust), called *conchas* (seashells), which are one of the favourites among the diverse universe of *pan dulce* in Mexico. As the name suggests, *conchas* are shaped like a clam shell; the crust can be coloured with food colouring and are typically flavoured with vanilla or chocolate.

Better known in Europe and elsewhere are chocolate shells. Scrumptiously rich and smooth marbled praline chocolate seashells were originally created by Belgian chocolatier Guylian in the shape of seashells. Today, these treats are sold in the shape of a few different shells as well as seahorses and shrimp. The company has supported the marine conservation institute Project Seahorse since 1993 and uses a seahorse as its company symbol. Many brands sell chocolate seashells, but the original seashells by Guylian come with a G sign on each piece as the company's trademark.

Shells are also found in the culinary world in shell-shaped plates and trays (for example, for salads, or small butter dishes in the shape of scallops). Some restaurants have real or imitation giant clam shells to present seafood. While they are beautiful, giant clams are endangered and protected internationally, unless they are imported from an aquaculture facility that raises them, like the one based in Micronesia. Most giant clam shells offered for sale on the Internet do not appear to have the proper paperwork that makes them legal.

Some games are played using shells as game pieces, for example the ancient game Mancala, which uses shells, beads, rocks, seeds or other objects, typically on depressions on a board. The game has two players who move their pieces in opposite directions on the board. The name Mancala is Arabic, meaning 'to move'; the game has many names

in different countries, but generally the rules are similar. The oldest examples of this game have been found in Israel, and date back to the second or third century.

Kai-awase, the Japanese memory game that dates back to the Heian Period (794–1185), was played by Japanese nobility, especially women. One to four players turn valves of the Asian hard clam (*Meretrix lusoria*) that have been painted with matching designs on the internal side of both valves of a pair. The objective is to find matching pairs and remember their position after turning the valves back to the unpainted external surfaces. The game was played with a full set of 360 pairs of shells, each pair painted with slightly different designs, traditionally taken from the eleventh-century work *The Tale of Genji*. The shells were selected to be about the same size and colour. The meaning of *kai-awase* translates to 'shell matching'. At the beginning of the game, the left and right valves would be arranged in two groups, and each player would turn two valves at a time, looking for a matching pair. Today full sets of *kai-awase* are found in museums, but smaller sets (the most basic is a single matching pair of clams) can be found for sale at museums and stores in Japan.

*Above: Belgian chocolatier Guylian invented chocolate seashells, which are now imitated by other brands. Opposite: Japanese shell-matching game (*Kai-awase*), 19th century, gold, colour and shell-white pigment on paper on shells.*

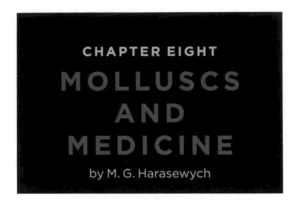

CHAPTER EIGHT

MOLLUSCS
AND
MEDICINE

by M. G. Harasewych

Molluscs have had impacts on human health, positive and negative, both real and imagined, from prehistoric times to the present day. The earliest as well as the largest contribution of molluscs to human health and well-being has been in the area of nutrition. Humans and their ancestors have been eating molluscs – marine, freshwater and terrestrial – for hundreds of thousands of years, and many molluscs, primarily bivalves, cephalopods and gastropods, continue to be staples of the human diet today.

Molluscs are one of the most commonly consumed types of seafood. They are nutrient-dense foods that tend to be low in calories, high in protein and rich in important micronutrients such as iron, zinc, selenium, manganese, vitamin B12 and antioxidants such as omega-3 fatty acids. They feature as a significant component of diets that are associated with decreased risks of cardiovascular diseases, various cancers and other maladies. Although consuming molluscs is generally considered beneficial to human health, some people may develop shellfish allergies, most commonly reacting to crustaceans. However, allergic reactions to all classes of molluscs, including bivalves (for example, clams, oysters, scallops, mussels), cephalopods (octopus and squid) and gastropods (limpets, conchs, abalone, whelks and winkles) have been reported. The reactions are most commonly associated with consuming molluscs, but in some cases can occur following occupational exposure.[1] Symptoms can range from mildly itchy skin to swelling of the face, lips, tongue and throat, as well as trouble breathing, nausea or even anaphylactic shock; the symptoms also generally become more severe after each exposure.

Consuming raw or undercooked oysters or other molluscs, especially in the summer, increases the risk of developing viral or bacterial infections, such as *Vibrio*, which occur naturally in coastal waters and can be concentrated in bivalves. Symptoms vary depending on the species of *Vibrio*, some causing mild symptoms such as diarrhoea and vomiting, others causing sepsis, which in extreme cases can lead to limb amputation and death.[2] Depending on the environment in which they were harvested, some molluscs can also accumulate heavy metals, such as mercury and cadmium, although at lower levels than fish.[3]

Several varieties of shellfish poisoning have been associated with consumption of bivalve molluscs that accumulate neurotoxins produced by algae, dinoflagellates and cyanobacteria they filter from the water during toxic algal blooms. The toxins accumulate in the tissues of the bivalve and can be passed to animals that eat them, ranging from sea otters and seals to humans. The primary toxin (saxitoxin) causing paralytic shellfish poisoning (PSP) is produced by dinoflagellates and is not destroyed by cooking. Some bivalves retain the toxin for weeks following the algal bloom; others can store the toxin for up to two years. Symptoms can range from mild (nausea, vomiting, diarrhoea, shortness of breath) to severe (death). Amnesic shellfish poisoning (ASP) is

*Blood fluke (*Schistosoma haematobium*), Egypt.*

caused by domoic acid, a neurotoxin produced by marine diatoms that is unaffected by cooking or freezing. Gastrointestinal symptoms occur before neurological symptoms, which may include short-term memory loss. Diarrhetic shellfish poisoning (DSP), a less dangerous form of shellfish poisoning, is caused by okadaic acid, which is also produced by dinoflagellates. Symptoms range from abdominal pains to diarrhoea and dehydration. Outbreaks of neurotoxic shellfish poisoning (NSP) may occur following toxic algal blooms, sometimes referred to as red tides, and are usually associated with the nearby death of fish and sea birds. Algal blooms have become more frequent and severe due to human activities resulting in eutrophication of coastal waters. Toxins produced by dinoflagellates may persist in bivalves for up to two months following the algal blooms. Symptoms range from nausea, vomiting and diarrhoea, to numbness and slurred speech, to respiratory distress. The poisoning of humans by shellfish following algal blooms has occurred throughout history. Native Americans learned to avoid consuming shellfish during periods of marine bioluminescence, which occurs during algal blooms.[4] Molluscs are among the foods prohibited by Jewish dietary laws (Kashrut) as they, like all invertebrates except locusts, are considered non-kosher. There are some differences among Muslim sects as to whether all or only some molluscs are permitted to be eaten (halal). Apart from individuals who develop food allergies, however, nearly all issues associated with consuming molluscs can be avoided if the molluscs are appropriately harvested and prepared.

Several groups of freshwater snails that occur throughout tropical and semi-tropical regions of the world may serve as intermediate hosts for parasitic worms that infect humans. Schistosomiasis, also referred to as snail fever or bilharzia, is the most prevalent of these diseases, affecting approximately 250 million people throughout the world in 2015, and causing an estimated 200,000 deaths per year. Among tropical parasitic diseases, it is second only to malaria in terms of economic impact.

The pathology is caused by worms that lodge in the blood vessels within the pelvis, the mesentery and the urinary tract. Adult worms lay eggs that pass through the walls of the vessels and are excreted with faeces and urine. These eggs must reach freshwater, where they break open to release minute, free-swimming larvae called miracidia that must reach an appropriate snail host to survive. These miracidia burrow into the snail through its skin, and then migrate to the digestive gland (liver) of the snail where they undergo two generations of sporocyst development and multiplication before producing forked-tailed larvae (cercariae) that are shed into the surrounding water from the ruptured tissues of the snail. When humans come into contact with contaminated water, the cercariae burrow through their skin into the peripheral capillary bed and are transported through the blood to the heart and lungs, and then to the portal vessels of the liver; here the blood flukes develop and grow before migrating to the blood venules within the pelvis, where they develop into adults and repeat the cycle. Adult worms can live for thirty years within their human host. There are three species of schistosome worms that infect humans, each with different snail intermediate hosts. *Schistosoma haematobium* is widespread throughout the Nile valley and occurs in Africa and portions of the Middle East. It infects the pelvic veins and affects the urinary system. *Schistosoma mansoni* also occurs throughout Africa, as well as in portions of South America and the Caribbean islands, where it was likely introduced by the slave trade; this worm infects the liver and intestines. *Schistosoma japonicum* is confined to eastern and southeastern Asia and adjacent islands including the Philippines. Pathology can vary depending on species, and severity of infection, primarily affecting people in rural and agricultural areas. It may include a rash, fever and anaemia, or inflammatory reactions affecting the intestinal wall, bladder, liver or brain.

Symptoms of schistosomiasis have been reported in early Egyptian records, and the disease has been discovered in Egyptian mummies from 1250 BCE. Napoleon's troops were infected with the disease during the Egyptian Campaign (1798–1801 CE). Construction of dams and irrigation canals in the second half of the twentieth century caused increases in the incidence of schistosomiasis.[5] The key to lasting preventive measures against the blood-fluke disease is improved sanitation and the elimination of the mollusc intermediate host.

There are numerous other snail-borne diseases caused by parasitic worms, for which freshwater molluscs serve as intermediate hosts. The lung fluke (*Paragonimus westermani*) in Asia and West Africa lodges in the lungs and produces often-fatal tuberculosis-like lesions. Eggs are coughed out of the lungs into the river, where they hatch as miracidia and penetrate a snail host, where they develop into rediae then cercariae that are shed into the water. These cercariae then infect freshwater crustaceans (crabs or crayfish) in which they encyst to form metacercariae, and humans are infected by eating the uncooked or undercooked crustaceans. Metacercaria then pass through the duodenal wall until they reach the lungs. Other parasitic flukes infect the liver, intestine and other organs, and may involve additional intermediate hosts (including fish) after completing life stages within freshwater molluscs. Still other parasitic diseases such as rat lung worm (*Angiostrongylus cantonensis*) can be transmitted by terrestrial snails and slugs. Rats are the common hosts for this nematode infection, but if humans consume the infected snails, the larvae migrate to and remain in the brain. Symptoms may include headaches, confusion, disorientation, coma and death.

It is difficult to imagine that a mollusc might be capable of physically injuring a human, unless one recalls the scene from Jules Verne's *Twenty Thousand Leagues Under the Sea* in which the crew of the submarine *Nautilus* do battle with a giant squid. As giant squid normally inhabit depths of more than 300 metres (1,000 ft), humans rarely encounter them, and if we do, it is most often as dead or dying specimens that have floated to the surface or washed up on beaches. Pliny the Elder reported a specimen over 9 metres (30 ft) long, and giant squid have been known to mariners since ancient times. The names Kraken (Norse), Lusca (Caribbean) and Scylla (Greek) have been attributed to legendary large cephalopods said to be capable of destroying sailing ships.

There have also been occasional apocryphal stories about giant clams causing the death of humans by closing upon the arms or feet of divers, leading to their drowning. However, giant clams are harmless, feeding on algae growing within their tissues. They move far too slowly to close their valves and trap a diver.

Several lineages of predatory molluscs have evolved venoms, and some have been documented to have caused human fatalities. Recent studies have shown that all octopuses and cuttlefish and some squid produce venom. The bite of the blue-ringed octopus that inhabits shallow waters in coral reefs of the Indo-West Pacific region can result in paralysis, nausea and death within minutes.

Among the gastropods, a highly evolved lineage of predatory snails belonging to the superfamily Conoidea (cone snails and their allies) has modified its teeth to form hollow, harpoon-like structures that it uses to impale its prey and inject them with paralytic toxins. These animals inhabit primarily tropical and semitropical seas throughout the world, tend to have colourful and attractive shells, and are popular with shell collectors. All cone snails are capable of 'stinging' humans if the living animals are handled, stabbing them with their hypodermic-needle-like teeth and injecting venom. The majority of species feed on worms or molluscs, and their stings are no worse than being stung by a bee or a wasp. However, certain species that have specialized to feed on fish produce toxins that are powerful enough to kill humans.

Many molluscan species have been used in traditional medicine in different parts of the world since antiquity. Shells, pearls, soft tissues and portions of entire animals have and continue to be formulated in various preparations for medicinal use to treat illnesses ranging from the common cold to gastrointestinal ailments, inflammation, cancer, tuberculosis and a host of other maladies. Hippocrates (460–370 BCE) prescribed the use of mucus from the commonly eaten land snail

*Alphonse de Neuville, 'Only one of his arms wriggled in the air,
brandishing the victim like a feather', illustration from Jules Verne,*
Twenty Thousand Leagues Under the Sea *(1871).*

(escargot, *Helix pomatia*) to treat a variety of ailments. Pliny (23–79 CE) recommended that snails reduced to a pulp could be used to treat burns and wounds, nosebleeds and stomach pain (when boiled and eaten with wine), and noted that snails from Africa are best, but they must be prepared in an uneven number. Claudius Galenus, or Galen (129–210 CE), one of the most accomplished medical researchers of antiquity, prescribed snails to treat various maladies.

Ambroise Paré, considered to be one of the foremost surgeons during the Renaissance and regarded as 'the father of modern surgery', prescribed snails as a treatment for anthrax. He also published a report of a giant amphibious snail, the Sarmatian snail, said to occur in the Baltic Sea. He commented that its flesh is 'very good and grateful meat', and its blood 'medicinable' for liver ailments, ulcerated lungs and leprosy.

Opposite: Diver trapped by a clam, drawing by Joseph M. Guerry from Wilburn Dowell Cobb, 'The Pearl of Allah', in Natural History, *XLIV/4 (November 1939).*
*Below: Blue-ringed octopus (*Hapalochaena *sp.), Lembeh, Indonesia.*

Preparations derived from molluscs continue to be used for many ailments such as ulcerated lungs and leprosy in parts of India, Africa and Latin America. Similarly, traditional Chinese medicine, still widespread in China and globally, includes molluscs as ingredients for treatments of numerous conditions, among them asthma, eczema, osteoarthritis and burns.[6] The continuous and widespread use of molluscs throughout history and into the modern era has prompted modern researchers to investigate in more detail the various components of such treatments and their therapeutic efficacy. Thus far, only a few hundred molluscan species have been studied, and over 1,000 natural products and metabolites have been isolated or tested for pharmacological activity. Proteins isolated from several gastropods and bivalves have been found to be effective against human microbial (for example, *Vibrio cholerae*, *Salmonella typhi*) and viral pathogens, including Human T-cell leukaemia virus type 1, Herpes simplex virus (HSV)-1 and human immunodeficiency virus type 1 (HIV-1). Other molluscan compounds have been shown to have potent anti-inflammatory activity and have led to the marketing of several nutraceuticals

*Illustration of the mythical Sarmatian sea snail (*Cochlea sarmatica*), from Ulisse Aldrovandi,* De reliquis animalibus exanguibus *(1606).*

(such as Lyprinol, Biolane and Cadalmin) to treat arthritis, asthma and other inflammatory conditions.

Many of the neurotoxins contained in the venom of several of the species of fish-eating cone snails have been studied in detail and found to consist of small peptide molecules. Some were found to have a rapid paralytic effect, others a powerful anaesthetic or analgesic effect. One, termed ω-conotoxin, was found to be extremely effective against severe pain (over 1,000 times more powerful than morphine). A synthetic version of this peptide, termed Ziconotide, has been approved for use by the u.s. Food and Drug Administration, for delivery by infusion into the cerebrospinal fluid for treatment of patients to manage severe chronic pain.

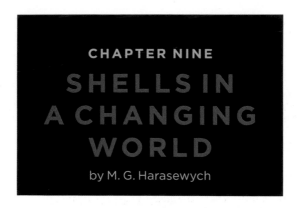

CHAPTER NINE

SHELLS IN A CHANGING WORLD

by M. G. Harasewych

Molluscs are among the oldest groups of living organisms, with several lineages dating back to Cambrian faunas (540 million years ago). They have survived five major mass extinctions (Ordovician, Devonian, Permian, Triassic-Jurassic and Cretaceous-Tertiary mass extinctions) over the past 450 million years, each of which caused the extinction of more than 75 per cent of all plants and animals living at the time, as well as other smaller or more localized extinction events (for example, the Middle Miocene disruption). Durations of these periods of extinction ranged from less than 20,000 years to roughly 200,000 years, while recoveries were on the order of millions to tens of millions of years. Following each extinction, the molluscs that survived proliferated to become the second most diverse group of animals living today (estimates approach 100,000 living species) and inhabit virtually all aquatic (both marine and freshwater) and terrestrial habitats.

In contrast, the lineage leading to humans appears to have originated approximately 500,000 years ago, with modern humans (*Homo sapiens sapiens*) being present on Earth for less than 50,000 years. In a fraction of this time, human activity has brought about the onset of a sixth major mass extinction, termed the Anthropocene extinction. The Anthropocene has been proposed as a geological epoch that is defined by the onset of substantial human impact on Earth's biota, climate, geology and chemistry. Various events have been proposed to denote the beginning of the Anthropocene, ranging from the first expansion of early humans out of Africa more than 45,000 years ago to the development of agriculture and animal domestication approximately 12,000 years ago, and the industrial revolutions that

*Giant clam (*Tridacna gigas*), Great Barrier Reef, Queensland.*

began in the late eighteenth and nineteenth centuries coinciding with the widespread use of machinery and the burning of fossil fuels, to the first nuclear explosion in 1945, all with major and accelerating impacts on the environment. The extinction rate during the Anthropocene is estimated to be 10 to 100 or more times higher than that in any previous mass extinction, and it is accelerating. Human impacts have caused the extinction of 322 species of terrestrial mammals since 1500, and possibly as many as 7.5 to 13 per cent (150,000 to 260,000) of the approximately 2 million known species as part of the 'Anthropocene Defaunation'.[1] Studies have projected that today's anthropogenic extinctions will diminish biodiversity for millions of years to come.[2]

The earliest effects humans had on molluscs were as foraging predators, consuming the more common species as food and using their shells for tools or for aesthetic purposes. With the rapid expansion of human populations, harvesting of edible molluscs increased quickly and was largely replaced by commercial fisheries and aquaculture for the most common species (for example, oysters, scallops, mussels, clams, abalone, conch, squid and many others). By 2018 world production included 17.7 million tonnes of molluscs as food and 26,000 tonnes of ornamental seashells and pearls, with molluscs representing more than 20 per cent of seafood consumed globally and half of the production of marine and coastal aquaculture.[3]

Many species of molluscs are being overfished, and regulations have been developed by governments at various levels, including local, national and international. The queen conch (*Aliger gigas*) continues to be fished throughout the Bahamas and various Caribbean islands but harvesting this species in Florida waters is prohibited by law. International trade in this species is regulated under the Convention on International Trade in Endangered Species of Wild Fauna and Flora (CITES) agreement. This species is presently not regarded as endangered in the Caribbean as a whole, but it *is* considered to be threatened in many areas due to overfishing and poaching. A recent study predicted that continued overfishing could lead to the extinction of this species in as little as ten years.[4]

Giant clams (Tridacninae), which include multiple species in the genera *Tridacna* and *Hippopus*, inhabit the tropical Indo-Pacific. They have been extensively collected not only for their meat but for their shells. Some species, such as *Tridacna gigas*, the largest of the giant clams, had been considered virtually extinct in parts of their range. All species are listed by CITES and international trade is prohibited. Breeding programmes have been in place since the 1980s to help restock wild populations throughout several Pacific Islands and portions of Japan and Southeast Asia.

Among the many other molluscs that had been commercially fished are five species of abalones (*Haliotis*) occurring off the coast of California. These populations have experienced severe declines since the 1960s. Two species, the white abalone (*Haliotis sorenseni*) and the black abalone (*Haliotis cracherodii*), the latter once one of the most abundant large intertidal molluscs along the California coast, have been listed as endangered species, and the commercial fishery for all abalones was closed in 1997.[5] Abalone also occur and are fished throughout the Pacific, and many countries have developed fishery management strategies to regulate their industries.

As humans proliferated throughout history, there was an increase in migration, colonization and trade between ever more distant populations. Humans brought with them, both intentionally and unintentionally, many species of animals, plants and microbes. As humans colonized and travelled between oceanic islands, they transported land snails with them from island to island. Examples include anthropogenic introductions of partulid tree snails throughout the Papua New Guinean archipelagos and throughout Polynesia as well as pre-Columbian and more modern introductions of cerionid land snails from Cuba to and throughout the Bahamian Islands.[6]

*Live channelled whelks (*Busycotypus canaliculatus*) for sale at a California seafood market, 2009.*

More recent introductions, such as that of the commercially fished whelk *Busycotypus canaliculatus*, a species endemic to the eastern United States, into the waters of San Francisco Bay in California in the 1940s, may have been intentional, or as a by-product of the intentional introduction of the eastern oyster (*Crassostrea virginica*). Several species that co-occur on oyster reefs along the east coast are now also found in San Francisco Bay but have not expanded their range beyond the bay. A recent survey reported 278 aquatic invasive species, including 46 molluscs, in California bays and estuaries, many with possible links to fishing vessels.[7]

Among the most widespread and well-studied of introductions of molluscs is that of the veined rapa whelk (*Rapana venosa*), a species native to the northwestern Pacific Ocean. Populations were introduced into the Black Sea in the 1940s, probably with oysters (*Crassostrea gigas*) from Japan. From there it dispersed into the Azov Sea, the Sea of Marmara, throughout the Mediterranean Sea, along the Atlantic coast of France and Spain and into the North Sea. Populations have also been discovered in Chesapeake Bay and the La Plata estuary between Uruguay and Argentina. Unlike *Busycotypus canaliculatus*, which produces juveniles that hatch from egg capsules as crawling young, *Rapana venosa* produces offspring that have a free-swimming larval stage, which lasts between two and four weeks before they metamorphose

Opposite: Veined rapa whelk (Rapana venosa), Black Sea. Above: Zebra mussels (Dreissena polymorpha), Missouri River, South Dakota.

into bottom-dwelling juveniles. Such free-swimming larvae in the ballast water of ships travelling from the Black Sea or the Mediterranean are thought to be the source of the introduction into Chesapeake Bay.[8]

Zebra mussels (*Dreissena polymorpha*) native to lakes and rivers draining into the Black and Caspian seas have been introduced throughout European and North American waterways, likely carried in the ballast water of ships, and continue to be spread by ships and pleasure boats throughout inland waterways. These small freshwater mussels (not closely related to marine mussels, Mytiloidea) are extremely prolific (densities of up to 10,000 mussels/m^2) and attach to hard substrates by byssal threads. In addition to ecological effects such as reducing dissolved oxygen, attaching to and killing native species of freshwater mussels and serving as a source of avian botulism, which has resulted in the deaths of many birds, they damage harbours and vessels. Water treatment and power plants are particularly affected, with free-swimming larvae entering the facilities and attaching to and clogging pipes, resulting in costs estimated at well over U.S.$500 million annually in the Great Lakes alone.

Once established, introduced species compete with native species for food and limited resources, and by predation. Three specimens of the giant East African land snail (*Lissachatina fulica*) released in a garden in Miami, Florida, in 1966 proliferated to more than 18,000 snails by 1972, when eradication efforts began. A specimen was again discovered in Miami in 2011, with more than 40,000 snails collected within the following six months. The probability of complete eradication of this species in Florida is considered low. This snail is known to carry a parasitic nematode (rat lungworm, *Angiostrongylus cantonensis*) that can cause meningitis in humans, dogs and other animals. The introduction of various species such as rats and lizards has led to the extinction or endangered status of many land snails, especially species endemic to islands.

Continued and accelerating growth in human populations has led to rapidly increasing use of resources, including land and water for agriculture and the development of cities, all of which have negative effects on the environment on a broad and ever-expanding scale. The impact of human populations on the biosphere has been staggering. A recent study has estimated that the biomass of humans (measured in gigatons of carbon) is an order of magnitude greater than that of all wild mammals combined, and the biomass of livestock (primarily cattle and pigs) is nearly twice that of humans.[9] Activity of early humans contributed to the extinction of Quaternary megafauna between 50,000 and 3,000 years ago, claiming about half of all large (that is, greater than 45 kilogram/100 lb) land mammals (among them woolly mammoths and giant sloths). This study also reports that the total biomass of plants today (and, by proxy, the total biomass on Earth) is approximately half of what it was at the start of human civilization, with crops cultivated by humans representing about 2 per cent of total present plant biomass. Another study reports that the total mass of human-made solid materials (anthropogenic mass) will exceed the total living biomass on Earth (about 1.1 trillion tonnes) in the year 2020 (± 6 years), and that the amount of anthropogenic mass doubles every twenty years. Each person on Earth, on average, produces their body weight in anthropogenic mass every week, and researchers predict that there will be more plastic than fish in the oceans of the world by 2050.[10]

Human activities such as deforestation, slash-and-burn agriculture and the construction of cities and roads, as well as other forms of habitat destruction, have brought and continue to bring about the extinction of countless species, primarily in terrestrial habitats, many before they have been 'discovered' by science and formally described (termed 'Centinelan extinctions' by the American biologist and naturalist E. O. Wilson).[11] The greatest and most immediate toll tends to be on species that are endemic to small areas such as islands. The largest proportion of Earth's biodiversity consists of undescribed species, the vast majority of which are invertebrates, which comprise more than 95 per cent of known animal species.[12] Estimates of the proportion of insects living today that are not yet described are on the order of 80 per cent,[13] while the roughly 85,000 species of mollusc that have thus far been described represent only about one-third of the species predicted to be living today, with 63 per cent living in the oceans, 29 per cent on land and 8 per cent in fresh water.[14]

The IUCN is composed of numerous government and civil organizations that work together in the field on nature conservation and promote the sustainable use of natural resources. This organization maintains a Red List of Threatened Species, which documents the conservation status and extinction risk of portions of the world's biota. Thus far, this list has assessed only a very small fraction of the more than 2.14 million species of plants and animals presently know to science, itself a small fraction of the estimated number of species. Of the more than 140,000 species that have been evaluated, over 28,000 are considered to be at risk of extinction due to human activity. This list is strongly biased towards mammals and birds, but includes data on 9,862 species of molluscs, listing 308 species as extinct, 19 extinct in the wild and 744 as critically endangered. The most affected species, according to the Red List, are those endemic to small areas and specific habitats, such as islands, rivers, streams and hot springs.[15] Less than 15 per cent of the species assessed are from marine environments. Although human activities affect ocean-dwelling species, so far, such activities have generally been limited to areas smaller than the range of the species. Oil spills, for example, can be catastrophic to areas in which they occur, but most species survive outside of those areas and eventually can repopulate affected areas when conditions improve. Some marine species can react to changing temperatures or conditions by altering their ranges. As waters warm, some species can adapt by altering or expanding their latitudinal ranges towards the poles in search of appropriate water temperatures; others may migrate into deeper, cooler water. Yet even molluscs inhabiting hydrothermal vents – small, isolated and remote hot springs along mid-ocean ridges on the deep ocean floor, often at depths of thousands of metres – have been added to the Red List. Of the 184 species that have been assessed, 39 species have been listed as critically endangered, 32 species as endangered and 43 as vulnerable, with vent faunas in the Indian Ocean at greatest risk due to deep-sea mining for rare minerals at hydrothermal vents and abyssal plains.[16]

In addition to the effects of habitat destruction, the onset of the Industrial Revolution has greatly accelerated the production and dispersal of numerous chemical compounds, wastes and toxins into the environment. These tend to have far broader and generally somewhat slower effects. Whether chemical waste is buried in dumps or originates as run-offs from mines or other industrial facilities, or from fertilizers and pesticides applied to crops, such water-soluble compounds permeate into the ground water, wash into streams and rivers and flow downstream into the ocean, affecting aquatic biotas along the way. Some – including spring snails that are limited to a very few springs in desert regions, freshwater bivalves of the family Unionidae and many species of snails that are each endemic to individual streams or drainages – can also be rapidly affected. Many such freshwater species have some of the highest extinction rates among molluscs. High levels of fertilizers, sewage and livestock waste reaching coastal

*The scaly-foot snail (*Chrysomallon squamiferum*), also known as the sea pangolin, one of the species at risk of extinction due to potential deep-sea mining by humans.*

waters have contributed to the increased frequency of harmful algal blooms, commonly called red tides. Various algae and dinoflagellates present in such blooms produce a variety of toxins that are harmful to many forms of marine life, commonly causing fish kills, in which large numbers of dead fish, as well as dolphins, turtles, birds and other marine animals, wash up on shore. Not all molluscs are susceptible to the toxins, and some filter-feeding bivalves can accumulate high levels of toxins in their tissues that may be fatal when eaten by humans, birds or other predators. Oil spills, another form of chemical pollution, have toxic effects on faunas inhabiting areas of the coast and ocean bottoms where they occur.

By far the most damaging, wide-ranging and long-term effects of human pollution are due to emissions of gases. Examples include the opening of the ozone hole over Antarctica due to ozone depletion in the upper atmosphere caused by the release of chlorofluorocarbons (CFCs) from coolants, aerosols and fire extinguishers into the atmosphere, permitting more ultraviolet rays to penetrate the atmosphere, causing increases in skin cancer, eye damage and crop damage. An international agreement to limit the release of CFCs in 1987 has led to the reduction in the size of the ozone hole with the prediction that it will return to its pre-1980 level by the mid-century.

The burning of fossil fuels for energy will undoubtedly be the major contributor to the Anthropocene mass extinction. Beginning in the 1920s, oil companies began adding tetraethyl lead to gasoline in order to enhance engine performance. Until this practice was banned, nearly a century later, lead microparticles from gasoline emissions had been settling into the soil, but they still tend to be resuspended into the atmosphere during hot, dry summer weather. Acid rain, produced when gases formed by burning sulphur-containing coal interact with water in the atmosphere, have substantial adverse effects on forests, bodies of freshwater and the organisms that live in them, as well as soils and microbes, in addition to being corrosive to metal and stone structures.

It is interesting to note that the Permian mass extinction, also referred to as the Great Dying, the most severe mass extinction in the history of the planet, was caused by rapidly rising carbon dioxide (CO_2) levels produced by volcanic eruptions in Siberia. These resulted in rapid global warming, expanding desert regions, rising sea levels, an increase in ocean temperatures of nearly 8.3°C (15°F), dissolution of CO_2 in ocean waters causing acidification and decreased dissolved oxygen levels in the sea. The Permian mass extinction occurred over a period of 200,000 years and caused the extinction of nearly 95 per cent of all species living at the time. A recent study has shown that the CO_2 emissions that caused the end-Permian extinction were roughly half of the emissions currently being produced on the planet.[17]

REFERENCES

Chapter 1 The Shell Makers

1 R. C. Brusca and G. J. Brusca, *Invertebrates*, 2nd edn (Sunderland, MA, 2003).

2 K. Muramatsu, J. Yamamoto, T. Abe, K. Sekiguchi, N. Hoshi and Y. Sakurai, 'Oceanic Squid Do Fly', *Marine Biology*, CLX/5 (2013), pp. 1171–5.

3 J. Leal and M. G. Harasewych, 'Deepest Atlantic Molluscs: Hadal Limpets (Mollusca: Gastropoda: Cocculiniformia) from the Northern Boundary of the Caribbean Plate with Descriptions of Two New Species', *Invertebrate Biology*, CXVIII/2 (1999), pp. 116–36.

4 D. L. Graf, 'Patterns of Freshwater Bivalve Global Diversity and the State of Phylogenetic Studies on the Unionoida, Sphaeriidae, and Cyrenidae', *American Malacological Bulletin*, XXXI/1 (2013), pp. 135–53.

5 J. A. Schneider, 'Phylogenetic Relationships and Morphological Evolution of the Subfamilies Clinocardiinae, Lymnocardiinae, Fraginae, and Tridacninae (Bivalvia: Cardiidae)', *Malacologia*, XL/1–2 (1998), pp. 321–73.

6 G. M. Barker, 'Gastropods on Land: Phylogeny, Diversity, and Adaptive Morphology', in *The Biology of Terrestrial Molluscs*, ed. G. M. Barker (Wallingford, New Zealand, 2001), pp. 1–146; E. E. Strong, O. Gargominy, W. F. Ponder and P. Bouchet, 'Global Diversity of Gastropods (Gastropoda; Mollusca) in Freshwater', Freshwater Animal Diversity Assessment, *Hydrobiologia*, DXCV/1 (2008), pp. 149–66.

7 See www.squid-world.com/colossal-squid, accessed 6 July 2022.

8 J. A. Mather, R. Anderson and J. B. Wood, *Octopus: The Ocean's Intelligent Invertebrate* (Portland, OR, 2010).

9 Katherine Harmon, 'Octopuses Gain Consciousness (According to Scientists' Declaration)', *Scientific American*, 21 August 2021.

10 F. Moretzsohn et al., 'Mollusca: Introduction', in *Gulf of Mexico: Origin, Waters, and Biota*, vol. 1: *Biodiversity*, ed. D. L. Felder and D. K. Camp (College Station, TX, 2009), pp. 559–64.

11 P. Bouchet, P. Lozouet, P. Maestrati and V. Heros, 'Assessing the Magnitude of Species Richness in Tropical Marine Environments: Exceptionally High Numbers of Molluscs at a New Caledonia Site', *Biological Journal of the Linnean Society*, LXXV/4 (2002), pp. 421–36.

12 E. A. Kay and R. Kawamoto, *Micromolluscan Assemblages in*

Mamala Bay, O'ahu, 1974–1982, Water Resources Research Center Report, University of Hawai'i at Manoa (Honolulu, 1983).

13 C.F.E. Roper and E. K. Shea, 'Unanswered Questions About the Giant Squid *Architeuthis* (Architeuthidae) Illustrate Our Incomplete Knowledge of Coleoid Cephalopods', *American Malacological Bulletin*, xxxi/1 (2013), pp. 109–22.

14 Henning Lemche, 'A New Living Deep-Sea Mollusc of the Cambro-Devonian Class Monoplacophora', *Nature*, clxxix/4956 (1957), pp. 413–16.

15 Philippe Bouchet, *Shells* (New York, 2007).

16 Leslie Crnkovic, message at '[Conch-L] Kibai Zufu, by Kenkado Kimura 1775 (*Pleurotomaria hirasei*)', at conch-l@listserv.uga.edu, 3 April 2015.

Chapter 2 Tribal Shell Use

1 J.C.A. Joordens et al., '*Homo erectus* at Trinil on Java Used Shells for Tool Production and Engraving', *Nature*, dxviii/7538 (2014), pp. 228–31.

2 David Campbell, 'Pushing Back the Dawn of Hominid Art', posted on the forum Conch-L, 4 December 2014, https://listserv.uga.edu.

3 Molecular evidence may provide different dates for the appearance of *Homo sapiens*, but its discussion is beyond the scope of this book.

4 Katherine Szabo, 'Prehistoric Shellfish Gathering', www.manandmollusc.net, 2002.

5 Avril Bourquin, www.manandmollusc. net, accessed 21 July 2022.

6 bbc One, 'Race Against the Tide, Risking Death Under Huge Blocks of Ice – Human Planet: Arctic' (2011), www.youtube.com.

7 Mitchell Clark, 'Some Basics on Shell Trumpets and Some Very Basics on How to Make Them', in *Sound Inventions: Selected Articles from Experimental Musical Instruments*, ed. B. Hopkin and S. Tewari (London, 1996), available online at www.furious.com, accessed 21 July 2022.

8 D. E. Bar-Yosef Mayer, B. Vandermeersch and O. Bar-Yosef, 'Shells and Ochre in Middle Paleolithic Qafzeh Cave, Israel: Indications for Modern Behavior', *Journal of Human Evolution*, lvi/3 (2009), pp. 307–14.

9 Paul Pettitt, *The Paleolithic Origins of Human Burial* (Abingdon-on-Thames, 2010).

10 Joseph Heller, *Sea Snails: A Natural History* (New York, 2015).

11 Robert R. Stieglitz, 'The Minoan Origin of Tyrian Purple', *Biblical Archaeologist*, lvii/1 (1994), pp. 46–54.

12 A. Hyatt Verrill, *Strange Sea Shells and Their Stories* (Boston, ma, 1936).

Chapter 3 Shells and Religion

1 D. E. Bar-Yosef Mayer, B. Vandermeersch and O. Bar-Yosef, 'Shells and Ochre in Middle Paleolithic Qafzeh Cave, Israel: Indications for Modern Behavior', *Journal of Human Evolution*, LVI/3 (2009), pp. 307–14; C. S. Henshilwood et al., 'A 100,000-Year-Old Ochre-Processing Workshop at Blombos Cave, South Africa', *Science*, CCCXXXIV/6053 (2011), pp. 219–22.
2 K. D. Rose, 'The Religious Use of *Turbinella pyrum* (Linnaeus), the Indian Chank', *The Nautilus*, LXXXVIII/1 (1974), pp. 1–5.
3 J. Hornell, 'The Sacred Chank of India; A Monograph of the Indian Conch (*Turbinella pyrum*)', *Madras Fisheries Bureau Bulletin*, 7 (1914), pp. 1–181, pls 1–18.
4 Harry Lee, 'Historical Notes on a Sinistral Sacred Chank: *Turbinella pyrum*', *American Conchologist*, XXXIX/2 (2011), pp. 28–9.
5 Ibid.
6 Tom Rice, 'What Is a Shell Worth?', *American Conchologist*, XXXIX/2 (2011), p. 27.
7 R. Beer, *The Handbook of Tibetan Buddhist Symbols* (Chicago, IL, 2003).
8 J. W. Jackson, 'The Aztec Moon-Cult and Its Relation to the Chank-Cult of India', *Manchester Memoirs*, LX/5, (1916), pp. 1–5.

Chapter 4 Picking Up Money on the Beach

1 P. Paraide, 'Formalizing Indigenous Number and Measurement Knowledge', *Journal of the Linguistic Society of Papua New Guinea*, XXXIII/2 (2015), pp. 1–15.
2 Ibid.
3 J. Hogendorn and M. Johnson, *The Shell Money of the Slave Trade* (Cambridge, 1986).
4 Peng Xinwei, *A Monetary History of China*, trans. Edward H. Kaplan (Bellingham, WA, 1993).
5 H. Scales, *Spirals in Time: The Secret Life and Curious Afterlife of Seashells* (London, 2015).
6 J. Hogendorn and M. Johnson, *The Shell Money of the Slave Trade* (Cambridge, 1986).
7 Ibid.
8 Ibid.; Scales, *Spirals in Time*; Avril Bourquin, '2. Trade Goods', www.manandmollusc.net, accessed 21 July 2022.
9 John McCabe, 'The Greeks', 2005, www.oysters.us.
10 S. Peter Dance, *Out of My Shell: A Diversion for Shell Lovers* (Sanibel Island, FL, 2005).
11 P. M. Rexford, 'Stamps Once Used as Money', *Washington Post*, Lifestyle, 7 November 1986, at www.washingtonpost.com.
12 T. Walker, 'Shells on Stamps', 2003, www.conchology.be.

Chapter 5 How the Cowrie Got Its Spots

1 S. Peter Dance, *Shell Collecting: An Illustrated History* (London, 1966).
2 F. Moretzsohn, 'Cypraeidae: How Well-Inventoried Is the Best-Known Seashell Family?', *American Malacological Bulletin*, XXXII/2 (2014), pp. 278–89.
3 E. Savazzi, 'The Colour Patterns of Cypraeid Gastropods', *Lethaia*, XXXI/1 (1998), pp. 15–27.
4 Hans Meinhardt, *The Algorithmic Beauty of Sea Shells* (Heidelberg, 1995).
5 M. Passamonti, 'The Family Cypraeidae (Gastropoda Cypraeoidea) an Unexpected Case of Neglected Animals', *Biodiversity Journal*, VI/1 (2015), pp. 449–66.
6 Clarence M. Burgess, *Cowries of the World* (Cape Town, 1985), p. xvii.
7 D. H. Bluestein and R. C. Anderson, 'Localization of Octopus Drill Holes on Cowries', *American Malacological Bulletin*, XXXIV/1 (2016), pp. 1–4.
8 H. Bradner and E. Alison Kay, 'An Atlas of Cowrie Radulae (Mollusca: Gastropoda: Cypraeoidea: Cypraeidae)', *The Festivus* Supplement, XXVIII (1996).
9 Savazzi, 'The Colour Patterns of Cypraeid Gastropods'.
10 Fabio Moretzsohn, 'Exploring Novel Taxonomic Character Sets in the Mollusca: The *Cribrarula cribraria* Complex (Gastropoda: Cypraeidae) as a Case Study', PhD diss., University of Hawai'i, Honolulu, 2003.

11 C. P. Meyer, 'Molecular Systematics of Cowries (Gastropoda: Cypraeidae) and Diversification Patterns in the Tropics', *Biological Journal of the Linnean Society*, LXXIX/3 (2003), pp. 401–59; C. P. Meyer, 'Towards Comprehensiveness: Increased Molecular Sampling with Cypraeidae and Its Phylogenetic Implications', *Malacologia*, XLVI (2004), pp. 127–56.
12 Fabio Moretzsohn and Serge Gofas, 'Cypraeidae: World Register of Marine Species' (2012), www.marinespecies.org.
13 For example, Alex Van Steen, 'Cowrie Shells: More than Simply "Shell Money"', 2 August 2013, https://climbcarstensz.wordpress.com; John Taylor and Jerry G. Walls, *Cowries* (Neptune City, NJ, 1975).
14 Felix Lorenz and Alex Hubert, *A Guide to Worldwide Cowries* (Wiesbaden, 1993), p. 571.
15 World Register of Marine Species, '*Sphaerocypraea incomparabilis* (Briano, 1993)', www.marinespecies.org, accessed 21 July 2021.

Chapter 6 Iridescent Beauty

1 Al Smith and Sarina Jepsen, 'Overlooked Gems: The Benefits of Freshwater Mussels', *Wings*, XXXI/2 (2008), pp. 14–19.
2 Neil H. Landman, Paula M Mikkelsen, Rüdiger Bieler and Bennet Bronson,

Pearls: A Natural History (New York, 2001).

3 Wilbur Dowell Cobb, 'The Pearl of Allah', *Natural History*, XLIV/4 (1939), pp. 197–202.

4 BBC News, '34 kg Pearl Found in Philippines "Is World's Biggest"', www.bbc.com, 24 August 2016.

Chapter 7 Shells in the Arts

1 K. J. Boss, 'References to Molluscan Taxa Introduced by Linnaeus in the Systema Naturae (1758, 1767)', *The Nautilus*, CII/3 (1988), pp. 115–22. Boss lists nearly ninety such works cited by Linnaeus.

2 S. J. Gould, 'Left Snails and Right Minds', *Natural History*, CIV/4 (1995), pp. 10–18; W. D. Allmon, 'The Evolution of Accuracy in Natural History Illustration: Reversal of Printed Illustrations of Snail and Crabs in Pre-Linnaean Works Suggests Indifference to Morphological Detail', *Archives of Natural History*, XXXIV/1 (2007), pp. 174–91.

3 S. Peter Dance, *Shell Collecting: An Illustrated History* (London, 1966).

4 Joseph Heller, *Sea Snails: A Natural History* (Heidelberg, 2015).

5 Linda Paik Moriarty, *Ni'ihau Shell Leis* (Honolulu, HI, 1986).

Chapter 8 Molluscs and Medicine

1 S. L. Taylor, 'Molluscan Shellfish Allergy', *Advances in Food and Nutrition Research*, 54 (2008), pp. 139–77; Y. Zhang, H. Matsuo and E. Morita, 'Cross-Reactivity Among Shrimp, Crab and Scallops in a Patient with a Seafood Allergy', *Journal of Dermatology*, XXXIII/3 (2006), pp. 174–7.

2 Centers for Disease Control and Prevention, 'Oysters and Vibriosis', www.cdc.gov, accessed 21 July 2022.

3 Wen-X. Wang and G. Lu, 'Heavy Metals in Bivalve Mollusks', in *Chemical Contaminants and Residues in Food*, ed. D. Schrenk and A. Cartus, 2nd edn (Duxford, 2017), pp. 553–94.

4 R. Munday and J. Reeve, 'Risk Assessment of Shellfish Toxins', *Toxins*, V/11 (2013), pp. 2109–37.

5 S. H. Sokolow et al., 'Nearly 400 Million People Are at Higher Risk of Schistosomiasis Because Dams Block the Migration of Snail-Eating River Prawns', *Philosophical Transactions of the Royal Society of London*, Series B: Biological Sciences, CCCLXXII/1722 (2017), available online at https://royalsocietypublishing.org.

6 T. B. Ahmad, L. Liu, M. Kotiw and K. Benkendorff, 'Review of Anti-Inflammatory, Immune-Modulatory and Wound Healing Properties of Molluscs', *Journal of Ethnopharmacology*, 210 (2018), pp. 156–78.

Chapter 9 Shells in a Changing World

1 R. H. Cowie, P. Bouchet and B. Fontaine, 'The Sixth Mass Extinction: Fact, Fiction, or Speculation?', *Biological Reviews of the Cambridge Philosophical Society*, XCVII/2 (2022), pp. 640–63; R. Dirzo et al., 'Defaunation in the Anthropocene', *Science*, CCCXLV/6195 (2014), pp. 401–6.

2 J. W. Kirchner and A. Weil, 'Delayed Biological Recovery from Extinctions through the Fossil Record', *Nature*, CCCIV/6774 (2000), pp. 177–80.

3 Food and Agriculture Organization of the United Nations (FAO), *The State of World Fisheries and Aquaculture 2020: Sustainability in Action* (Rome, 2020).

4 A. W. Stoner, M. H. Davis and A. S. Kough, 'Relationships between Fishing Pressure and Stock Structure in Queen Conch (*Lobatus gigas*) Populations: Synthesis of Long-Term Surveys and Evidence for Overfishing in the Bahamas', *Reviews in Fisheries Science and Aquaculture*, XXVII/1 (2019), pp. 51–71.

5 W. S. Leet, C. M. Dewees, R. Klingbeil and E. J. Larson, *California's Living Marine Resources: A Status Report* (Sacramento, CA, 2001), pp. 11–16.

6 D. Ó Foighil, T. Lee and J. Slapcinsky, 'Prehistoric Anthropogenic Introduction of Partulid Tree Snails in Papua New Guinean Archipelagos', *Journal of Biogeography*, XXXVIII/8 (2011),

pp. 1625–32; T. Lee et al., 'Prehistoric Inter-Archipelago Trading of Polynesian Tree Snails Leaves a Conservation Legacy', *Proceedings of the Royal Society*, Series B, CCLXXIV/1627 (2007), pp. 2907–14; C. J. Maynard, Appendix to 'Contributions to the History of Cerionidae with Descriptions of Many New Species and Notes on Evolution in Birds and Plants', *Records of Walks and Talks with Nature*, X/1–2 (1919–26), pp. 1–218, pls 5–43.

7 A. N. Cohen and J. T. Carlton, *Nonindigenous Aquatic Species in a United States Estuary: A Case Study of the Biological Invasions of the San Francisco Bay and Delta* (Washington, DC, 1995), pp. 1–49.

8 National Estuarine and Marine Exotic Species Information System (NEMESIS), 'Rapana venosa' (2022), https://invasions.si.edu.

9 Y. M. Bar-On, R. Phillips and R. Milo, 'The Biomass Distribution on Earth', *Proceedings of the National Academy of Science USA*, CXV/25 (2018), pp. 6506–11.

10 E. Elhacham et al., 'Global Human-Made Mass Exceeds All Living Biomass', *Nature*, DLXXXVIII/7838 (2020), pp. 442–4.

11 E. O. Wilson, *The Diversity of Life* (Cambridge, MA, 1992), pp. 1–424.

12 B. R. Scheffers, L. N. Joppa, S. L. Pimm and W. F. Laurance, 'What We Know and Don't Know About Earth's Missing Biodiversity', *Trends in Ecology and Evolution*, XXVII/9 (2012), pp. 501–10.

13 N. E. Stork, 'How Many Species of Insects and Other Terrestrial Arthropods Are There on Earth?', *Annual Review of Entomology*, LXIII (2018), pp. 31–45.

14 A. D. Chapman, *Number of Living Species in Australia and the World*, 2nd edn (Canberra, 2009); P. Bouchet, S. Bary, V. Héros and G. Marani, 'How Many Species of Molluscs Are There in the World's Oceans, and Who Is Going to Describe Them?', *Mémoires du Muséum national d'Histoire naturelle*, 208 (2016), pp. 9–24.

15 International Union for Conservation of Nature (IUCN), 'The IUCN Red List of Threatened Species, 2020–2021' (2020), www.iucnredlist.org.

16 E. A. Thomas et al., 'A Global Red List for Hydrothermal Vent Molluscs', *Frontiers in Marine Science* (2021), available online at www.frontiersin.org.

17 Y. Cui et al., 'Massive and Rapid Predominantly Volcanic CO_2 Emission During the End-Permian Mass Extinction', *Proceedings of the National Academy of Sciences of the USA*, CXVIII/37 (2021), available online at www.pnas.org.

SELECT BIBLIOGRAPHY

Shell collecting has been a very popular pastime for many centuries, and there have been a great many books written on the subject of molluscs, ranging from those intended for the general reader to more detailed scientific reports for specialists and researchers. With the advent of the Internet, an increasing number of websites have become available intended for a wide range of audiences, many spanning multiple topics related to molluscs, shells and shell collecting.

Abbott, R. Tucker, *American Seashells*, 2nd edn (New York, 1974)
—, *Compendium of Landshells* (Melbourne, FL, 1989)
—, *Kingdom of the Seashell* (Melbourne, FL, 1993)
—, and S. Peter Dance, *Compendium of Seashells*, 2nd edn (Melbourne, FL, 1982)
Barnett, Cynthia, *The Sound of the Sea: Seashells and the Fate of the Ocean* (New York, 2021)
Bouchet, Philippe, and Gilles Mermet, *Shells* (New York, 2008)
Dance, S. Peter, *Rare Shells* (London, 1969)
—, *A History of Shell Collecting* (Leiden, 1986)
—, *Shells: The Clearest Recognition Guide Available* (London, 2002)
—, *Out of My Shell: A Diversion for Shell Lovers* (Sanibel Island, FL, 2005)
Fearer Safer, Jane, and Frances McLaughlin Gill, *Spirals from the Sea: An Anthropological Look at Shells* (New York, 1982)
Fyodorov, A., and H. Yakovlev, *Mollusks: Morphology, Behavior and Ecology* (New York, 2012)
Harasewych, M. G., *Shells, Jewels from the Sea* (New York, 1989)
—, and Fabio Moretzsohn, *The Book of Shells: A Life-Size Guide to Identifying and Classifying Six Hundred Seashells* (Chicago, IL, 2010)
Landman, Neil H., Paula M. Mikkelsen, Rüdiger Bieler and Bennet Bronson, *Pearls: A Natural History* (New York, 2001)
Lydeard, C., and K. S. Cummings, *Freshwater Mollusks of the World: A Distribution Atlas* (Baltimore, MD, 2019)
Matsukuma, Akihiko, Takashi Okutani and Tadashige Habe, *World Seashells of Rarity and Beauty* (Tokyo, 1991)
Pisor, Don, *Sea and Land Shells of the Don Pisor Collection: Color, Form, Shape* (Hackenheim, 2015)
Ponder, W. F., D. R. Lindberg and J. M. Ponder, *Biology and Evolution*

of the Mollusca, vol. I (Abingdon, 2019), vol. II (Abingdon, 2020)

Stix, Hugh, Marguerite Stix and R. Tucker Abbott, *The Shell, Five Hundred Million Years of Inspired Design* (New York, 1968)

Sturm, C. F., T. A. Pearce and A. Valdes, *The Mollusks: A Guide to Their Study, Collection, and Preservation* (Irvine, CA, 2006)

Vermeij, Geerat J., *A Natural History of Shells* (Princeton, NJ, 1936)

Verrill, A. Hyatt, *Strange Sea Shells and Their Stories* (Boston, MA, 1936)

ASSOCIATIONS AND WEBSITES

American Malacological Society
A society for researchers working with
molluscs
www.malacological.org

Broward Shell Club
One of several shell clubs featuring
websites that provide links to information
for shell collectors
https://browardshellclub.org

Conchbooks
A publisher and dealer specializing in
books on shells, both recently published
and antiquarian
www.conchbooks.de

Conchologists of America (COA)
A society for shell enthusiasts at all levels
of interest
www.conchologistsofamerica.org

Conchology
An extensive website with links to many
topics involving shells, including price
lists for shell books and specimen shells
www.conchology.be

Freshwater Mollusk Conservation
Society
https://molluskconservation.org

Jacksonville Shell Club
Website providing links and information
on a wide variety of topics involving shells
http://jaxshells.org

Mollia
Sources of information for malacologists
www.ucmp.berkeley.edu/mologis/mollia.
html

National Shellfisheries Association
A society for studies on shellfish,
including molluscs
www.shellfish.org

Seashell and Mollusc Links
Website with numerous links to national
and international sites featuring various
aspects of collecting and studying shells
www.petersseashells.com/shelllinks.html

Unitas Malacologica
An international society to further the
study of molluscs
www.unitasmalacologica.org

LARGE COLLECTIONS OF SHELLS

Natural history museums are present in many larger cities, and many museums, including some university museums, have research collections of shells, in addition to exhibits open to the public.

Eastern USA

Department of Malacology, Museum of Comparative Zoology
Harvard University
26 Oxford Street,
Cambridge, MA 02138
https://mcz.harvard.edu/malacology

Department of Living Invertebrates
American Museum of Natural History
Central Park West at 79th Street
New York, NY 10024
www.amnh.org/research/invertebrate-zoology

Department of Malacology
The Academy of Natural Sciences of Drexel University
19th and Parkway
Philadelphia, PA 19103
https://ansp.org/research/systematics-evolution/malacology

Mollusk Department
Delaware Museum of Natural History

P. O. Box 3937
Wilmington, DE 19807-0937
www.delmnh.org/collections-research/mollusks

Department of Invertebrate Zoology
National Museum of Natural History
Smithsonian Institution
10th and Constitution Avenue, NW
Washington, DC 20013-7012
http://invertebrates.si.edu

Department of Natural History
Florida Museum of Natural History
3125 Hull Road
Gainesville, FL 32611
www.floridamuseum.ufl.edu/nhdept

Bailey-Matthews National Shell Museum
3075 Sanibel Captiva Road
Sanibel, FL 33957
www.shellmuseum.org

Central USA

Division of Invertebrates
Field Museum of Natural History
1400 S DuSable Lake Shore Drive
Chicago, IL 60605
www.fieldmuseum.org/science/research/area/invertebrates

Illinois Natural History Survey
Forbes Natural History Building
University of Illinois
1816 South Oak Street
Champaign, IL 61820
https://mollusk.inhs.illinois.edu/research

Division of Mollusks
University of Michigan Museum of
Zoology
1109 Geddes Avenue
Ann Arbor, MI 48109-1079
https://lsa.umich.edu/ummz/mollusks.
html

Department of Mollusks
Carnegie Museum of Natural History
4400 Forbes Avenue
Pittsburgh, PA 15213
https://carnegiemnh.org/research/
mollusks-malacology

Houston Museum of Natural Science
5555 Hermann Park Drive
Houston, TX 77030
www.hmns.org/exhibits/permanent-
exhibitions/malacology

Western USA/Hawaiian Islands

The Burke Museum
4300 15th Street NE
Seattle, WA
www.burkemuseum.org/collections-and-
research/biology/malacology

Department of Invertebrate Zoology and
Geology
California Academy of Sciences, Golden
Gate Park
55 Music Concourse Drive
San Francisco, CA 94118
www.calacademy.org/scientists/
invertebrate-zoology-geology

Department of Invertebrate Zoology
Santa Barbara Museum of Natural History
2559 Puesta del Sol Road
Santa Barbara, CA 93105
www.sbnature.org/collections-research/
invertebrates

Department of Malacology
Natural History Museum of Los Angeles
County
900 Exposition Boulevard
Los Angeles, CA 90007
https://nhm.org/research-collections/
departments-and-programs/malacology

Department of Marine Invertebrates
San Diego Natural History Museum
1788 El Prado
San Diego, CA 92101
www.sdnhm.org/science/marine-
invertebrates

Department of Malacology
Bishop Museum
1525 Bernice St
Honolulu, HI 96817
www.bishopmuseum.org/malacology

United Kingdom
Department of Invertebrates
The Natural History Museum
Cromwell Road
London SW7 5BD
England, UK
www.nhm.ac.uk/our-science/
departments-and-staff/life-sciences/
invertebrates.html

Department of Natural Sciences
Amgueddfa Cymru – National Museum
Wales
Cathays Park
Cardiff CF10 3NP
Wales, UK
https://museum.wales/biology/
biodiversity/mollusca

Natural Sciences Department
National Museum of Scotland
Chambers Street
Edinburgh EH1 1JF
Scotland, UK
www.nms.ac.uk/collections-research/
collections-departments/natural-sciences/
collections/invertebrates/marine-
invertebrates-collection

Europe

Muséum national d'Histoire naturelle
57 Rue Cuvier
75005 Paris
France
www.mnhn.fr/en

Muséum d'histoire naturelle
Route de Malagnou 1
1208 Genève
Switzerland
www.geneve.ch/en/museum-natural-
history

Koninklijk Belgisch Instituut voor
Natuurwetenschappen
Vautierstraat, 29
1000 Brussel
Belgium
www.naturalsciences.be/nl/science/
collection_page/518

Naturalis Biodiversity Center
Darwinweg 2
2333 CR Leiden
Netherlands
www.naturalis.nl/en/collection

Museum für Naturkunde
Invalidenstrasse 43
10115 Berlin
Germany
www.museumfuernaturkunde.berlin/en/
science/molluscs

Russia

Zoological Institute of the Russian
Academy of Sciences
Universitetskaya nab. 1
St Petersburg 199034
Russian Federation
www.zin.ru/sitemap_en.html

South Africa

Iziko South African Museum
25 Queen Victoria Street
Cape Town, South Africa
www.iziko.org.za/museums/south-
african-museum

KwaZulu-Natal Museum
237 Jabu Ndlovu Street
Pietermaritzburg, KwaZulu-Natal 3201
South Africa
www.nmsa.org.za/mollusca.html

India

Bombay Natural History Society
Hornbill House
Opp. Lion Gate
Shaheed Bhagat Singh Road
Fort
Mumbai 400 001
Maharashtra
India
www.bnhs.org

China

Shanghai Natural History Museum
399 Sha Hai Guan Lu
Jingan Qu
Shanghai Shi
China 200041
www.snhm.org.cn/cpjc_eg/indcx.htm

Japan

National Museum of Nature and Science
7-20 Ueno Park
Taito-kui
Tokyo 110-8718
Japan
www.kahaku.go.jp/english/research/
department/zoology/index.html

Australia

Australian Museum
1 William Street
Sydney NSW 2010
Australia
https://australian.museum/learn/
collections/natural-science/malacology

Western Australian Museum
Collections and Research Centre
49 Kew Street
Welshpool
Western Australia 6106
Australia
https://museum.wa.gov.au/online-
collections/users/aquatic-zoology

New Zealand

Museum of New Zealand Te Papa
Tongarewa
55 Cable Street
Wellington 6011
New Zealand
https://collections.tepapa.govt.nz/topic/442

PHOTO ACKNOWLEDGEMENTS

The authors and publishers wish to express their thanks to the below sources of illustrative material and/or permission to reproduce it. Some locations of artworks are also given below, in the interest of brevity:

Alamy Stock Photo: pp. 104 (Kirn Vintage Stock), 121 (Alvis Upitis); from Ulisse Aldrovandi, *De reliquis animalibus exanguibus* (Bologna, 1606), photo Biblioteka Narodowa, Warsaw: p. 136; Brooklyn Museum, New York: p. 48 (CC BY 3.0); The Cleveland Museum of Art, OH: pp. 42, 124; Cooper Hewitt, Smithsonian Design Museum, New York: p. 96 (*right*); courtesy The Elizabeth Taylor Trust, photo © 2011 Christie's Images Limited: p. 105; photo Maryann Flick: p. 59; Flickr: pp. 35 (photo Marc Carlson, CC BY 2.0 – Woolaroc Museum, near Bartlesville, OK), 58 (photo Marko Kudjerski, CC BY 2.0), 85 (*left*; photo Laszlo Ilyes, CC BY 2.0); Gallerie degli Uffizi, Florence: p. 112; photo Jennifer Granneman/FWC: p. 21; from Ernst Haeckel, *Kunstformen der Natur* (Leipzig and Vienna, 1904), photo Library of Congress, Washington, DC: p. 9; photos M. G. Harasewych, Smithsonian Collections: pp. 99, 113; Heritage Auctions, HA.com: pp. 95 (*left* and *right*), 123; Iconographia Zoologica – Special Collections of the University of Amsterdam: p. 18; iStock.com: pp. 12 (izanbar), 103 (brytta); The J. Paul Getty Museum, Los Angeles: p. 92; photo courtesy Scott and Jeanette Johnson: p. 84; from Charles C. Jones Jr, *Antiquities of the Southern Indians, Particularly of the Georgia Tribes* (New York, 1873), photo University of Illinois Urbana-Champaign: p. 41; from L.-C. Kiener, *Spécies général et iconographie des coquilles vivantes*, vol. IV (Paris, 1843), photo Smithsonian Libraries, Washington, DC: p. 14; photo Kristin La Flamme: p. 37 (*bottom*); Library of Congress, Prints and Photographs Division, Washington, DC: pp. 38, 49; from Martin Lister, *Historiæ sive synopsis methodicæ conchyliorum et tabularum anatomicarum* (Oxford, 1770), photo M. G. Harasewych: p. 109; photos Wim Lustenhouwer (VU University Amsterdam), courtesy José Joordens and Naturalis Biodiversity Center (RGM. DUB.1006.f): pp. 26, 27; The Metropolitan Museum of Art, New York: pp. 96 (*left*), 115; photo courtesy Dr Chris Meyer, Smithsonian Institution: p. 77; Minneapolis Institute of Art, MN: pp. 43, 127; photos Fabio Moretzsohn: pp. 7, 13 (*top right*), 22, 34, 36, 52 (*right*), 65, 72 (*top and bottom*), 74 (*top and bottom*), 78, 80, 82, 102; Musée du Louvre, Paris: pp. 107,

INDEX

Page numbers in *italics* refer to illustrations